# The Lean Games and Simulations Book

# Second Edition

Participative Games and Simulations for
Learning Lean:
An Instruction Book for Facilitators
covering
process and people in manufacturing and
service

John Bicheno
University of Buckingham
Lean Enterprise Unit

PICSIE Books, January 2015

# The Lean Games and Simulations Book
Second Edition

Published by
**PICSIE BOOKS**

How to order:
Telephone and Fax 01280 815023
Web site: www.picsie.co.uk

Copies of handouts in the appendix are also available in PDF format by email: bichenojohn@me.com at a cost of £20 plus VAT

Copyright: PICSIE Books 2010, 2015
Published January 2015
ISBN 978-0-9568307-2-2
British Library Cataloguing-in-Publication Data
A catalogue record for this book is available from the British Library

# CONTENTS

| | |
|---|---|
| **Summary Table of Games** | **7** |
| **Kata** | **10** |
| **AAR** | **12** |
| **A3 Improvement** | **15** |
| **Airplane Games** | **19** |
| JIT Basic Game | 20 |
| Cell Layouts Types Airplane | 25 |
| Bucket Brigade Line Balance | 28 |
| TWI Job Breakdown Exercise | 33 |
| **Squares Games** | **35** |
| Systems Thinking | 36 |
| 5S | 39 |
| Changeover Reduction | 41 |
| **Batch Dice Games** | **44** |
| Muda Muri Mura | 45 |
| Basic Batch Dice Game | 51 |
| Process Variation Reduction | 56 |
| Reducing Input Variation | 58 |
| Kanban: good and not so good | 60 |
| Flexible Working | 63 |
| Bottleneck and Overcapacity | 66 |
| Drum Buffer Rope | 68 |
| CONWIP | 71 |
| OEE Implications and Limitations | 74 |
| DBR with Backlog and Constraint Loops | 77 |
| **Single Piece Dice Games** | **80** |
| Basic Single Piece Flow | 81 |
| DBR and CONWIP | 85 |
| Critical WIP and Best Case Flow | 86 |
| Kanban Game | 89 |
| CONWIP and DBR Advanced | 92 |
| **Cut and Fill Games** | **96** |
| Unbalanced Push | 99 |
| Line Balance with Push and One Piece Flow | 100 |
| Line Balance with Kanban and One Piece Flow | 103 |
| Line Balance with CONWIP | 106 |
| Bucket Brigade Balancing | 107 |
| **5S Numbers** | **109** |
| **Cell Design and Balancing** | **112** |
| **The Paper Work Cell (Office Kaizen)** | **116** |
| **Area Supervisor Email Exercise** | **120** |
| **Call Centre, Failure Demand and Targets** | **123** |
| **TPM Checklist** | **130** |

Spot the Rot                                    **133**
TV News Game                                    **135**
Multitasking                                    **138**
Simultaneous Projects                           **140**
Lean Design                                     **143**
The Happy Pig                                   **147**
The Standards Debate                            **149**
The Supplier Partnership Game                   **152**
Go See: Genchi Genbutsu                         **158**
Targets and Measures                            **160**
Diversity and the Wisdom of Crowds              **163**


**Appendix**                                    **166**
    A3                                          167
    Airplane Games                              176
    Square Games                                203
    Dice Games                                  223
    Kanban Games                                234
    Cut & Fill Games                            245
    5S                                          259
    Manufacturing Cell                          268
    Paperwork Office                            272
    Service Call Centre Operations              283
    Spot the Rot                                294
    TV Game                                     295
    Multitasking                                309
    Lean Design                                 320
    Happy Pig                                   330
    Standards Debate                            336
    Go See                                      337
    Targets & Measures                          339

# Introduction to First Edition

There are many excellent books about the tools, the people, the culture, the leadership, and the implementation of Lean. I hope that The Lean Toolbox, 4th Edition, falls into some of these categories!

However, Lean is essentially about participation. Of course you cannot learn about Lean only by reading books. You need to "Go to the Gemba" and learn by doing.

That is where games come in. Most people working in Lean are visual, participative learners. Seeing is believing. Although no substitute for experiencing the Lean process, games have great impact and often bring out far more than the nominal focus of the game, to include multi-dimensional factors such as team working, decision-making, and practical problem solving. These are aspects that few if any books can cover.

In fact more experienced Lean practitioners know that the real reason for using various Lean tools is not for the headline name of the tool, but what it leads on to. The real reason for management participation in kaizen events is to learn that improvement skills are widespread throughout all levels of the organisation, the real reason for kanban is to expose problems, the real reason for 5S is to create stability, the real reason for A3 is to develop people, and so on. So it is with the games in this book.

This book is not intended to be 'stand alone'. You will want to use it side-by-side with such books as *The Lean Toolbox, Lean Thinking*, and *Leading the Lean Enterprise Transformation.*

Nor is the intention that the games will be played in turn, beginning to end.

But going through the games will give great insight, even to experienced Lean practitioners. There is always something new to learn about Lean! And, experienced games players know that even after running a game several times, new insights arise out of the comments of players.

None of the games in this book requires extensive or expensive materials. Dice, coloured pens, A4 paper, pairs of scissors, are generally all that is required. More extensive (and expensive!) games are available through the PICSIE Books website.

In the Appendix you will find games cards that will need to be copied and distributed to players. These figures are also available on Powerpoint or pdf format by emailing piciebook@btinternet.com.

# Introduction to Second Edition

The first edition was called the Lean Games Book. In this second edition several simulations and games have been added, and many have been revised and updated.

The first edition was taken up by many organisations in several countries. The interest generated was a motivation to develop and test new games and to revise some of those from the first edition.

There are 13 new games or simulations, some of them extensive.

I take a broad view of 'Lean' and therefore include games that can be connected with six sigma, Factory Physics, Theory of Constraints, TWI, Deming philosophy, Lean accounting, and psychology – as well as more conventional Lean areas. Many of these are required for successful Lean implementation.

'Kata' and AAR (After Action Reviews) can be incorporated with many of the games, old and new. Likewise doing an A3 exercise alongside the games will give practice in structured problem solving. These three new games or exercises are the first three in the book. Combining one of these with other games will extend the time, but will help to embed the learning for participants and will help the instructor to improve facilitation skills.

Some of the new games and simulations require more materials than in the first edition, but these are invariably inexpensive everyday items.

Many of the games have been tested during the MSc in Lean Enterprise at Buckingham. Others have been used on short courses and on in-house programs.

Thanks are due to MSc Lean students (all of whom are practising managers) for their suggestions and ideas.

I think you will enjoy facilitating the games. Let me know how you get on.


Buckingham Lean Enterprise Unit (BLEU)
University of Buckingham
August 2014


Copies of handouts in the appendix are also available in PDF format by email:
bichenojohn@me.com at a cost of £20 plus VAT

**Recommended Accompanying Books**

**On Lean:**

James Womack and Daniel Jones, *Lean Thinking*, revised edition, Free Press, 2003

George Koenigsaecker, *Leading the Lean Enterprise Transformation,* Second edition, CRC Press, 2012

Mike Rother, *Toyota Kata*, McGraw Hill, 2009

Niklas Modig and Par Ahlstrom, *This is Lean*, Rheologica, 2012

**Technical Detail in Support of Manufacturing Games:**

John Bicheno and Matthias Holweg, *The Lean Toolbox*, 4th edition, PICSIE, 2009

**Technical Detail in Support of Service Games:**

John Bicheno, *The Service Systems Toolbox,* PICSIE, 2012

**On Games Facilitation:**

Tom Justice and David Jamieson, *The Facilitator's Fieldbook*, Second edition, AmaCom, 2006. (Much more than you will need here, but many useful ideas.)

# Summary Table of Games

| Game Name | Area | Keywords | Manuf or Service | Approx. Time | Players |
|---|---|---|---|---|---|
| Kata | Stage by stage improvement and learning | Continuous improvement, learning, PDCA, kaizen, small step improvement | Manuf & Service ? | Many games extended by 30 mins | As for the particular game |
| AAR | Post game learning experience | Continuous improvement, Learning, Kaizen, Team review | Manuf & Service | Many games extended by 15 minutes | As for the particular game |
| A3 Improvement | Systematic problem solving. Focus | A3 methodology, Problem solving, Focus, Pareto analysis, '6 Honest Serving Men', Quality, Mentoring | Manuf & Service | 90 minutes | 15 |
| JIT Airplane Basic Game | Overview: waste and flow | Waste, Layout, Lead time, Kanban, Pull, Postponement, One piece flow, Quality, Self and successive checks. Mapping possibilities. | Manuf Service? | 2 hours | 8 to 20 |
| Cell Layout types Airplane Game | Layout and cell alternatives. | Cell design, Layout, job enrichment, productivity, line, cell, cross-functional working. | Manuf & Service ? | 1.5 hours | 8, 16, 24 |
| Bucket brigade Line Balance | Line balancing | Line Balance, Cells, Productivity, Job variety, Job timing | Manuf | 1 hour | 6 plus on-lookers |
| TWI job breakdown Exercise | Job analysis, beginning standard work | Job breakdown, Standard Work, SOP, Training within Industry, Job Instruction | Manuf & Service ? | 1 hour | Up to say 15 |
| Systems Thinking Squares | Holistic problem solving - right brain, vs. linear left brain. | Systems, co-operation, implementation, teamwork, resistance to change, communication, parts and wholes. | Manuf & Service | 30 minutes | 5, 10, 15, 20 |
| 5S Squares | Housekeeping and standard work | 5S, Waste, Housekeeping, Productivity, Variation reduction, Standards | Manuf & Service | 1 hour | Any; teams of 3 to 5 |
| Changeover Reduction Squares | Changeover reduction | Changeover, setup, SMED, waste, batch sizing. Mapping possibilities. | Manuf & Service ? | 40 to 90 mins | Any; Teams of 4 to 6 |
| Muda Muri Mura Dice | Critical variables of Lean | Waste, Variation, Capacity Utilization, Queuing, Lead Time. | Manuf & Service | 30+ minutes | 8+ in pairs. Max 36 |
| Basic Batch Dice Game | Throughput and capacity | Line capacity, variation and dependent events, TOC, Laws of Factory Physics, line balance | Manuf & Service | 40 minutes | 6 to 11 per game: 33 max. |
| Batch Dice: Process Variation Reduction | Reason for 5S, std work, 6 sigma | Throughput, variation, WIP inventory, | Manuf & Service | 20 minutes but a follow on | As for basic dice |
| Batch Dice: Reducing Input Variation | Importance of Stabilizing input | Throughput variation. | Manuf & Service | 30 minutes but a follow on | As for basic dice |
| Batch Dice: Kanban: Good and not so Good | Kanban, pull and variation | Kanban and stage by stage pull limitations. | Manuf & Service | 30 minutes | As for basic dice |
| Batch Dice: Flexible Working | Capacity and Flexibility. | Flexible working, flexible labour, target output, input and output, capacity | Manuf & Service | 1 hour | As for basic dice |

| | | | | | |
|---|---|---|---|---|---|
| Batch Dice: Bottleneck and Overcapacity | Scheduling. Bottleneck effects | Overcapacity and bottlenecks An introduction to the Drum Buffer Rope and CONWIP games. Overproduction | Manuf & Service? | 30 minutes but a follow on | As for basic dice |
| Batch Dice: Drum Buffer Rope (DBR) | Scheduling. Controlling flow | DBR operation, WIP reduction with throughput. TOC. Feedback | Manuf & Service? | 30 minutes | As for basic dice |
| Batch Dice: CONWIP | Scheduling. Controlling flow | CONWIP vs. DBR. Multi-stage kanban alternative. Feedback control. | Manuf & Service? | 20 minutes follow on from DBR | As for basic dice |
| Batch Dice: OEE Implications and Limitations | TPM | Availability, quality, speed, interactions. OEE calculation. Limitations of OEE | Manuf | 40 minutes | As for basic dice |
| Batch Dice: DBR with backlog and constraint loops | Scheduling. Multi stage control | Two types of rope or pull signals and buffers. Communication, feedback. | Manuf & Service | 40 minutes | As for basic dice |
| Basic Single Piece Flow Game | Range of Dice Games but single piece | As for dice games above but with single piece rather than batch production. | Manuf & Service? | 45 minutes per game | 5 to 15 |
| Single Piece Flow: DBR and CONWIP | Scheduling Controlling flow | DBR and CONWIP with single piece flow. Multi stage pull | Manuf | 40 minutes | 5 to 15 |
| Single Piece Flow: Critical WIP and Best Case Flow | Inventory levels for a line with variation | Critical WIP, Best Case Performance, Bottlenecks, Variation, Factory Physics | Manuf | 45 to 75 minutes | 5 to 15 |
| Kanban Game | Operation of kanban loops, no of kanbans | Kanban, pull, pull signals, Kanban loop, Supermarket | Manuf, some service? | 75 minutes for all | Teams of 5; max 4 |
| CONWIP & DBR Advanced | Scheduling, bottlenecks | DBR and CONWIP with varying order size and priorities | Manuf | 75 mins | Teams of 5 |
| Cut and Fill: Unbalanced Push | Line Balance | Multi product cell design, Productivity, Efficiency, Activity timing | Manuf & Service?? | 30 minutes | 11 min; max 22 |
| Cut and Fill: Line Balance with push and one piece flow | Complex Line Balance, Cell Design | Multi product cell design, Line balance, Productivity, Efficiency. | Manuf & Service?? | 60 minutes to 1.5 hours | 11 min; max 22 |
| Cut and Fill: Line Balance with Kanban Pull and One Piece Flow | Complex Line Balance, Cell Design | Multi product cell design, Productivity, Efficiency. Kanban, Mixed model | Manuf & Service?? | 40 to 60 minutes | 11 min; max 22 |
| Cut and Fill: Line Balance with CONWIP | Complex CONWIP | Mixed model, CONWIP | Manuf & Service?? | 40 minutes | 11 min; max 22 |
| Cut and Fill: Bucket Brigade Balancing | Complex Line Balance | Bucket Brigade line balance, multi product | Manuf & Service?? | 40 minutes | 11 min; max 22 |
| 5S Numbers | Housekeeping productivity quality | 5S, Waste, Housekeeping, Productivity, Variation reduction, Standards. | Manuf & Service | 20 minutes | Any |
| Cell Design and Balancing | Taking out micro waste, redesign of cell | Cell design, layout, takt time, line balance, Standard operations chart, and waste removal. | Manuf | 1.5 hours | Groups of 3 to 5 up to total 20 |
| The Paperwork Cell (Office Kaizen) | Taking out micro waste, redesign of cell | As for Cell Design but in a Warehouse or Despatch environment | Service | 1.5 hours | Groups of 3 to 5 up to total 20 |

The Lean Games and Simulations Book

| | | | | | |
|---|---|---|---|---|---|
| Area Supervisor E Mail Exercise | Dealing and prioritising communications daily work | Implementation, leader standard work, priority, span of control, delegation, system, flow | Manuf & Service? | 1 hour | Groups of 3 to 5 up to total 20 |
| Call Centre, Failure Demand and Targets | Call centre operation, capacity, failure demand | Call Centre, Service Centre, Failure demand, abandoned calls, rework, capacity, queues. | Service | 80 mins | Teams of 9 to 12; one or two teams |
| TPM Check list | Learning to see self maintenance opportunity | TPM, Gemba, vision, checks, inspection, operator centred maintenance, autonomous maintenance, FMEA | Manuf | 30 minutes to 1 hour | Groups of 3 to 5 up to total 20 |
| Spot the Rot | Learning to see waste | Visual management. Spotting wastes | Manuf & Service? | 30 minutes | Any |
| TV News Game | Lessons in running a design office and projects | Lean Design, Lean Project management, Concurrent Design, Cross Functional working, Office Waste, Lead time reduction. | Manuf & Service | 2.5 hours | 10 min, up to 20 |
| Multitasking | The problems of multitasking | Multitasking, productivity, waste, projects | Service & Manuf? | 30 minutes | Teams of 5; max 20 |
| Simultaneous Projects | Alternatives to running simultaneous projects | Lean design, multi project management, uncertainty | Manuf and Service | 90 minutes | Teams of 4; max 16 |
| Lean Design | Competitive teams build artifacts and compare against customer satisfaction | Lean design, Project management, customer value, project cost tradeoffs, 3P (Production Preparation Process), Understanding customers | Manuf and Service | 2.5 to 3.5 hours | Teams of 6; max 24 |
| The Happy Pig | Standard work | Standard work, SOP. | Manuf & Service | 20 minutes | Any |
| The Standards Debate | Setting standards? | Call centres, Front office operations, failure demand | Service | 30 minutes | Any |
| The Supplier Partnership Game | Supplier partnership | Partnership, Supply chain, negotiation, Game theory, Trust | Manuf & Service? | 1 hour | Groups of 3 to 5 up to total 20 |
| Go see; Genchi Genbutsu | What do you notice? | Go see, Direct Observation, Learning to see, Going to the Gemba | Manuf & Service | 30 minutes | Any |
| Targets and Measures | Use and abuse of targets and measures. | Inappropriate targets, Motivation, Understanding variation, SPC, control limits, Regression to the mean, Deming | Manuf and Service | 45 minutes | 12 |
| Diversity and Wisdom of Crowds | When the group is better than an expert | Participation, Respect, Knowledge, Judgment | Manuf and Service | 30 minutes | Any |

# Kata

**Key Areas**

Continuous improvement, learning, PDCA, kaizen, small step improvement

**Manufacturing or Service?**

Both

**Overview**

This exercise is an activity that can be added on to many of the games in the book. It will aid understanding, add reinforcement as well as giving practice in the Kata approach.

The word Kata comes from Karate. This is a series of standard movements (for example Kata 1 to Kata 5) that are carried out from beginner to multi-dan black belt. They become habit or the natural way things are done. Apparently 'pathways' become established in the brain. So 'Toyota Kata' becomes the standard, embedded way to improve or do kaizen.

'Kata' is an approach introduced by Mike Rother in his influential book 'Toyota Kata'. It makes the powerful case for small, everyday improvements and learning from them. It is a powerful implementation methodology. Not doing 'Kata' could help to explain the failure of many Lean initiatives – trying to take too big a leap.

The exercise also uses Rother's '5 Questions', which are central to the 'improvement kata' methodology.

**Summary of Learning Points**

- Establishing the Target Condition makes Lean implementation more effective and less threatening
- The difference between a 'target' and a 'target condition'
- Traps to avoid in setting Target Conditions
- Kata as a way to institutionalise PDCA
- Kata helps fix the relationship between actions and their results, thereby aiding learning

**Approximate Time**

Incorporating 'kata' into a game could extend the game by 30+ minutes.

**Description**

- Kata involves limiting players' choice to one at a time. The selection should be framed as a 'target condition'.
- At the start the Target must be clearly stated. This is the outcome or goal to be aimed at, in the long run. The steps get one ultimately to the Target
- The next step should be an action statement of how the process should operate in the next round and should

- be specific (not vague, not an outcome, nor a countermeasure)
  - involve an expectation or forecast
  - be one action, not several
  - move the game towards a desired outcome or next target
  - ask what do we need to do next? Improve the process or problem solve? – but take only one step at a time.
- Use the 5 Steps of Kata (from Rother):
  1. What is the target condition?
  2. What is the actual condition now?
  3. What obstacles are preventing you from reaching the target condition? Which one are you addressing now?
  4. What is the next step?
  5. Go and see what we have learned from taking the step.
- The next step should be thought of as an experiment. This is the link with the PDCA cycle.
- Carry out the step.
- Compare what was achieved with what was expected.
- Make the lessons learned in the step explicit.
- Decide on the next step or target condition.

## Players

- Any number. Where used with games having several teams, each team should do a Kata exercise each round.

## Player Instructions

- This is an instructor-facilitated exercise, at least initially. Teams will eventually become self-assured in the Kata methodology and will then need little or no facilitation.
- Use the steps in the Description, particularly the 5 Steps.

## Forms, Graphs, Tables, Typical Results

- A flip chart and pens for each team is recommended.

## Instructor notes and discussion

- This is powerful! If you are going to do this you should certainly read Mike Rother, *Toyota Kata*, McGraw Hill, 2010

## Equipment

None

# After Action Review (AAR)

Key Areas:  Continuous Improvement, Learning, Kaizen, Team review

**Manufacturing or Service?**

Both

**Overview**

After Action Review (AAR) has become a highly successful improvement and learning methodology adopted by the US Army, and now practised routinely. It is a form of kaizen, with similarities to PDCA. Many organisations have attempted to copy the procedure with varying degrees of success.

It is made up of four apparently simple steps for the group to discuss after playing a game or doing any real world improvement activity. It is NOT an evaluation activity but a learning experience.

This book of course has many games for the facilitator to try. A great thing to do is to follow each game with an AAR. That way, you will not only embed the learning but also (for the players) learn and practice a hugely successful methodology, and (for the instructor) learn how to run the game better next time!

An AAR is best conducted immediately after the game or project. This essential point is sometimes not followed, resulting in loss of effectiveness.

It is a superb way to break down hierarchy.

**Summary of Learning Points**

- Give practice in the four steps of AAR
- AAR embeds learning
- AAR is a learning experience for both group members and instructor
- It is not an evaluation
- A good facilitator is required, but this can be learned and practised
- AAR should be a non-threatening way of getting excellent feedback.

**Approximate Time**

20 minutes

**Description**

The instructor (or an experienced third party) will act as facilitator.

A good idea is to divide a flip chart into four regions, each with the heading of the four steps below. Then record as the group discusses.

The four steps are:

1. Objective: What did we set out to do? What was planned?
2. Reality: What actually happened? This is not judgmental or an evaluation; it is simply the facts about what happened at each stage of the game or project.
3. Learning: Why did it happen that way? What went right and wrong? What did not meet expectations? What went well? Again, stick to the facts; it should not be personal. No blame. This is a learning step so ask what caused the results to turn out the way they did.
4. Next time: What should be changed next time: planning, processes, behaviours, what should be kept.
5. Or players, whatcould they do diffently at their workplace?

## Players

Any number, but all should participate.

## Player Instructions

An AAR has been shown to be an extremely effective learning mechanism. It is also superb team building. The exercise will give you the opportunity to experience the ARR process before you 'do it for real' back in your work place.

- The instructor should facilitate.
- Be open and honest.
- Avoid blame and judgment.
- Stick to the facts.
- Give others a chance to participate.
- Be positive
- Make sure that you the learning points and 'take aways'.
- Make sure that changes are incorporated next time.

## Forms, Graphs, Tables, Typical Results

None

## Instructor notes and discussion

- If this is the first time you are participating in an AAR, and you are not a skilled facilitator, try to find and use a skilled facilitator.
- The AAR should be done immediately after playing a game.
- An AAR is not a one-off activity. It should be done after EVERY game, if at all possible.
- Explain that an AAR is NOT AN EVALUATION. It is a learning experience.
- VERY IMPORTANT: Take time over the first two steps. DO NOT skip immediately to step 3! According to the US Army approximately one third of total time should be spent on the first two steps.
- Of course, if the group has mixed views on the Objective, this is a significant learning for you!
- STICK TO THE FACTS. DO NOT ALLOW PERSONAL REMARKS or evaluation.
- Let the group discuss; don't put your views forward.

- Remain unbiased.
- Watch out for domination by individuals. Ensure that everyone has a say.
- If there are personal issues, deal with them after the AAR in a private session.
- Hierarchy needs to be downplayed.
- Encourage openness and honesty.
- Peter Senge (MIT Systems) says that many AARs fail because they become a 'sterile technique' done without real commitment or interest, with the results accumulating in corporate documentation.
- You will learn to be a much better game instructor!
- Finally, if no action is taken as a result of the AAR, it is waste of time.
- Point out similarities with the Deming view: '85% of problems lie with the system, only 15% with the person.'
- Point out similarities with the Toyota view of not blaming the person, or the TWI statement 'If the worker hasn't learned the instructor hasn't taught'.
- Point out similarities with PDCA / Plan Do Study Act.

## Equipment

- Flip chart and pens

## References and Further Reading

- David Garvin, *Learning in Action*, Harvard, 2000
- Website on AAR from US Army

# A3 Improvement

## Key Areas

A3 methodology, Problem solving, Focus, Pareto analysis, '6 Honest Serving Men', Quality, Mentoring

## Manufacturing or Service?

Both. The problem situation used in the game is in service, but the principles apply to both manufacturing and service.

## Acknowledgement

The original concept for the game was developed by Justin Watts with several MSc Lean student groups. This is an extended version that attempts to simulate an actual problem discovery process. Of course, A3 has become a popular problem solving approach, but is not always done well.

## Overview

The Problem: there is an apparent problem with revenue in a coffee shop in a small town. The A3 format for problem solving is used to analyse the situation and uncover the problem.

The game begins by giving participants a brief description of the problem situation.

If there is time, teams are asked to brainstorm out possible problems and causes using a fishbone diagram. The 'top issue' is chosen. This is used for later comparison with the analysis carried out using A3 methodology.

The A3 format is then briefly described. It is a structured methodology, following PDCA.

Players must uncover the situation by analysis and asking questions. If questions are not asked, information is not given. Thus the game is not a case study where all information is given up front.

A comparison is made with the fishbone study. (Is it a 'fishbone' or a 'wishbone'?)

## Summary of Learning Points

- Taking too quick a view of the situation can lead to incorrect problem definition and solving the wrong problem. This is also known as 'premature convergence'.
- The A3 template is a powerful and structured methodology.
- The root problem is uncovered layer by layer.
- Kipling analysis (6 Honest Serving Mean) is very useful.
- A3 is a focused methodology that is first focused on identifying and addressing the most relevant issue, not trying to tackle a host of problems.
- A3 is a form of PDCA (Plan Do Check Act): think about all four stages.
- Mentoring assists with A3. John Shook says 'It takes two to A3'.

**Approximate Time**

90 minutes if fishbone brainstorming is included. One hour to 90 minutes if only the A3 part is played.

**Description**

- Before the game starts, the 'answer' strips should be set out on a table away from players' vision. Strips should be grouped into piles of similar answers, with at least the same number of strips as there are teams. If there is space available the strips could be arranged in sets – one set per team.
- Players are set in teams of 4 to 6. Each player is given a copy of the one-page briefing.
- For teams of 5 or 6, one player in each team should be appointed as mentor. This person gets the Mentor Question sheet, and is asked not to join in with the problem analysis but periodically to try to keep the team on track by asking appropriate questions from the Mentor Question Sheet.
- If there is time available ask the teams to brainstorm out possible causes of 'the problem' using fishbone analysis. This should be done on flipcharts. The 'big issue', which should be tackled first, should be identified by circling the text on the fishbone.
- A3 format templates are distributed. Of course, A3 size is ideal – one per group.
- A brief explanation of the A3 template is given by the instructor.
- Teams are asked to proceed with the analysis, taking it as far as implementation recommendations.
- A deadline should be given; 45 mins or an hour is suitable, depending on time available.
- Teams work through to completion or to the deadline time.
- Teams report back. If the fishbone option was played, a comparison should be made.
- A reflection and comment session concludes the exercise.

**Players**

Several simultaneous teams each of 4 players. If the full group size is 15 or more, split the group into teams of 4 players and one mentor per team

**Player Instructions**

- The group is split into teams.
- Each member of each team gets a Coffee Shop Briefing sheet.
- If asked by the instructor, each team should brainstorm out possible causes using a fishbone chart with the 6 M's : men / people; machine; method; materials; measures; mother nature / environment), and select the most likely cause.
- Teams are briefed by the instructor on the use of A3 problem solving.
- Each team is given two copies of a blank A3 template. The extra copy is because teams sometimes wish to re-start.
- Teams then attempt to 'solve' / improve / come up with countermeasures for the Coffee Shop.
- The instructor will tell the teams the stop time for the exercise.
- Teams must ask the manager for specific information they require. This is a 'pull' system.
- No information is pushed to the teams.
- Teams should fill in the A3 sheet template as they proceed.

- At the appointed time, each team presents its findings.
- If a team has a mentor, the mentor should use the Mentor Guidance Sheet to assist the team. The mentor should not join in with the actual analysis.

## Instructor notes and discussion

- The instructor plays the role of the coffee shop manager who can answer questions and provide information if asked. This must be explained to the players.
- The teams are told that they must ask questions to the manager (instructor). If they don't ask, they don't get.
- An important part of the game is to get players to focus on the specific main issue. There are several less important issues, but players need to use the pareto process to home in on the 'big one' and then, in turn, explore possible causes again using pareto.
- The main problem is students serving cold cappuccino in the morning. Students don't realise that warm milk is required for cappuccino. This in turn stems from poor training. But note that cappuccino is specifically made with hot milk and froth.
- Note that there are far fewer problems with filter coffee and espresso. Focus first on the big problem.
- Likewise there are issues with layout, supply and maybe signage. But these are for later.
- The ladies seem to know how to make cappuccino; the students do not. Focus on the students!
- The instructor gives answers to teams by providing the appropriate strips. Only one strip at a time. The simulation would pretend that the manager would have to go and find out the information requested by, for example, a survey or looking at records.
- Sometimes a strip will not be available to answer a specific question. In this case the instructor can improvise by making up a suitable comment. See the briefing about the coffee shop.
- Note that in this coffee shop Cappuccino is served, not Latte. They are basically the same thing, differing in the milk and foam added. Don't get into a discussion on this!
- Maintenance of coffee machines is by an outside contractor who provides a regular, good service.
- Cakes are served in the morning. There has never been an issue with these.
- Don't make up wild scenarios. It is OK to say, 'I don't know the answer to that question'.
- 'Go see' is an important part of the A3 process. If teams ask about this, give them the particular process step strip that they ask for OR give them the Layout strip. Only give them the information strips that they ask for. Do not give out both process strips unless they specifically ask for both.
- TWI-type job instruction would be a good recommendation – with main steps, key points, and reasons for key points. However, a full-blown JI course is not required!
- Appoint someone as trainer? Also check on the effectiveness of the training and retention amongst trained staff.

## /continued

**Equipment**

- **The Information Sheets, A3 templates, and information strips are to be found in the Appendix**
- Make copies of Briefing Document for each team.
- Information Sheet cards will be need to be copied, cut out, and placed in piles.
- Handout A3 blank
- Handout 'Model Solution'

**References and Further Reading**

- Durward Sobek and Art Smalley, *Understanding A3 Thinking*, CRC Press, 2008
- John Shook, *Managing To Learn*, Lean Enterprise Institute, 2008

# The Airplane Set of Games

Building a paper airplane is a fun way to learn many aspects of Lean. The Airplane set of games is made up of:

- A JIT / Lean Airplane Basic game
- Cell Layout Types Airplane Game
- A 'Bucket Brigade' Line-Balance game
- An exercise on TWI-type job instruction, including the job breakdown sheet.

In each game the same basic airplane is used. See the standard airplane folding sheets.

The basic design is probably similar, or exactly like, the paper airplanes you made in junior school.

The instructor should practice making a few airplanes before running any game. Accuracy of folding is important – particularly the initial folds that should be precisely aligned. Casual or slipshod folding will lead to poor airplanes that are difficult to fold later on, and that will not fly well.

The instructor should have a few completed airplanes on display for all airplane games.

**The Order of Playing the Games:**

- The JIT Airplane Basic Game can, of course, be played independently. This is a powerful way to show beginners the effects of pull against push. Not suitable for Kata.
- Generally, play the Bucket Brigade game before playing the Cell Layout Types Airplane Game. Knowing about line balance will be useful.
- The TWI game should follow either the Bucket Brigade Game or the Cell Layout Types Game.

# JIT Airplane Basic Game

This is a 'classic' Lean/ JIT game. Most JIT games are based on the original HP 'Stockless Production' video first shown externally in 1982 at the APICS Conference.

**Key Areas**

Waste, Overproduction, push versus pull Layout, Lead time, Kanban, Pull, Postponement, One piece flow, Quality, Self and successive checks.

**Manufacturing or Service?**

Manufacturing, but pull principles are useful in many service situations.

**Overview**

The game is played over 3 rounds. Paper airplanes are built. The first round is push, followed by pull 3 and then pull 1. Measures are taken each round. Dramatic improvements take place in lead time, WIP Inventory, Space. Quality issues become clear instead of hidden. Supplier relations and problem solving attitudes can be discussed.

**Summary of Learning Points**

- Introducing pull eliminates overproduction – the prime waste category.
- Pull reduces WIP inventory and makes quality problems easier and faster to detect.
- As overproduction and WIP decrease, so space can be reduced.

**Players**

7 minimum but timing, recording and supplier jobs can be found for three or four more players. Others observe.

**Approximate Time**

1 hour

**Description**

In this game, players do not make decisions. So, it is more of a demonstration than a game. Players build a standard paper airplane. The first round is a push system with players simply pushing planes along to the next player. Some time into the first round, the instructor introduces a marker sheet to the first station. The instructor also changes the pen colour of the player adding decals. This represents a defect. Packing envelopes are used by the last player. Part way into the round, a different size packing envelope is introduced. Measures are taken when the marker plane is completed. When the envelope size is changed a little drama is played out.

In the second round a kanban square of 3 is placed between workstations. Players make in batches of 3 using a pull system. Again, the instructor introduces a marker sheet and changes pen colour. Measures are taken.

The third round uses a kanban square of one. Measures are taken and written on a flip chart.

**Playing Instructions**

### First Round: Push

- The instructor will demonstrate how to build a paper airplane.
- Each player (1 to 8) will be given the appropriate Airplane instruction sheet.
- Players should be arranged along a table or bench, preferably straight but otherwise L shaped, with about 1 m between players.
- The final player is supplied with a stack of A4 size envelopes. Airplanes must be inserted 3 per envelope.
- Get the players to practice making planes. Part completed planes are left as work in process (WIP). Then start the round.
- When starting, have a pile of plain A4 sheets in front of the first player. The 20th sheet, however, should be a coloured (or marked) sheet. This sheet serves merely to measure the lead time, and should be treated no differently by players.
- When the round starts each player does his task as soon as part-completed planes become available from the previous player, and then pushes the plane to the next station.
- After a dozen or so planes have been started, the instructor will change the colour of the pen used by Player 5 adding the decals. The player should merely continue as before. (Blue Pens are changed to Red Pens.)
- When the first player begins work on the coloured marker sheet, a person should start a timer.
- When the marker sheet is completed by the final player, stop the timer and record the lead time.
- The instructor will then pause the game to record the number of defectives that have to be reworked. A defective is a plane with wrong coloured decals.
- The instructor will comment, saying that 'Quality problems tend to be hidden'.
- The instructor will also comment on the space being used. A suitable measure could be the number of tables used, or the length of the line from start to finish.

### First Round instructions for the final player and for a 'Supervisor' player

- Soon after the marker sheet has reached the final player, and the lead time has been recorded, the instructor will remove the A4 envelopes and substitute A5 envelopes. When an A5 size envelope is supplied, the final player must neatly fold the planes exactly in half and place a paperclip on each plane to prevent it from unfolding.
- Of course, this takes longer. The player should be asked to complain loudly.
- When the player complains, the Supervisor player should come along to the complaining player and ask what the trouble is. The final player explains, loudly, that he or she does not have sufficient time to do this extra work.
- The Supervisor then tells the player, again loudly, that he or she must just quit complaining, get on with the job, because the company 'pays for performance'.
- The final player continues at a faster rate but neglects to add the paperclips. (Brief the player to do this before the game starts.)

**Second Round: Pull 3**

- The second round introduces pull.
- A kanban square is placed between each workstation. Each 'square' is simply an A4 sheet, landscape view, divided into three sections using two vertical lines.
- Have a pile of plain A4 sheets in front of the first player. The 20th sheet, however, should be a coloured (or marked) sheet. This sheet serves merely to measure the lead time, and should be treated no differently by players.
- Start the game with three partly built up planes on each square. Before the round starts, the players should construct these, filling the squares from end to beginning but not making any completed airplanes.
- The operating rule is that each player makes in batches of three. For each player, the authorisation to start making is that the square after the player is empty. If it is full, the player is not authorised to make. Also, each player must wait for the square before the station to have three planes available, because the batch size is three.
- Players take all three partly completed planes from the square between the player and the previous player, and when completed place the planes on the next square.
- After a dozen or so planes have been started, the instructor will change the colour of the pen used by Player 5 adding the decals. The player should merely continue as before. (Blue Pens are changed to Red Pens.)
- When the first player begins work on the coloured marker sheet, a person should start a timer.
- When the marker sheet is completed by the final player, stop the timer and record the lead time.
- The instructor will then pause the game to record the number of defectives that have to be reworked. A defective is a plane with wrong coloured decals.
- The instructor will also comment on the space being used. A suitable measure could be the number of tables used, or the length of the line from start to finish.
- In this round envelopes sizes are not changed.
- Record the measures on the flip chart.

**Third Round: Pull 1 (or 'One Piece Flow')**

- A kanban square is placed between each workstation. Each 'square' is simply an A4 sheet, landscape view.
- Because the inter-player kanban squares now need to hold only one plane, the space, or tables, used can be reduced.
- Have a pile of plain A4 sheets in front of the first player. The 20th sheet, however, should be a coloured (or marked) sheet. This sheet serves merely to measure the lead time, and should be treated no differently by players.
- Start the game with one partly built up plane on each square. Before the round starts, the players should construct these, filling the squares from end to beginning but not making any completed airplanes.
- The operating rule is that each player makes in batches of one. For each player, the authorisation to start making is that the square after the player is empty. If it is full, the player is not authorised to make.
- Players take a partly completed planes from the square between the player and the previous player, and when completed place the plane on the next square.

- After a dozen or so planes have been started, the instructor will change the colour of the pen used by Player 5 adding the decals. The player should merely continue as before. (Blue Pens are changed to Red Pens.)
- When the first player begins work on the coloured marker sheet, a person should start a timer.
- When the marker sheet is completed by the final player, stop the timer and record the lead time.
- The instructor will then pause the game to record the number of defectives that have to be reworked. A defective is a plane with wrong coloured decals.
- The instructor will also comment on the space being used. A suitable measure could be the number of tables used, or the length of the line from start to finish.
- Soon after the marker sheet has reached the final player, and the lead time has been recorded, the instructor will remove the A4 envelopes and substitute A5 envelopes. When an A5 size envelope is supplied, the final player must neatly fold the planes exactly in half and place a paperclip on each plane to prevent it from unfolding.
- Of course, this takes longer. The player must complain loudly as in the first round.
- When the player complains, the Supervisor player should come along to the complaining player and ask what the trouble is. The final player explains, loudly, that he or she does not have sufficient time to do this extra work.
- But this time, the Supervisor tells the player, again loudly, that the problem will be sorted out immediately with the supplier, but in the meantime continue to use the A5 envelopes and to use paperclips.
- Note that, because kanban squares are being used, this will slow the line but quality should not be affected.

**Players**

- 7 players (Select a 'loud' player for the final station)
- A player to do timing
- A player to record results.
- A 'Supervisor' player, who must be familiar with Supervisor duties, as above.

## Materials

- A sheaf of A4 size plain paper
- Use the Airplane Game shhets (Operations 1 to 7 only)
- Three coloured A4 sheets (or clearly marked up plain sheets)
- A timer (smart phone may be suitable)
- Red and Blue felt tipped pens
- Paperclips
- Stapler
- Game Sheets (Note: only use one colour at a time: Blue then Red)
- Flip chart for results
- Note: Arrange the layout in a line or L shape; do not use the job shop type layout in the diagram. This is for the next game.

## Measures and Typical Results

|  | Round 1 Push | Round 2 Pull 3 | Round 3 Pull 1 |
|---|---|---|---|
| Lead time | 8:40 | 6:25 | 2:30 |
| Space | 4 tables | 3 tables? | 2 tables |
| WIP | 48 | 27 | 18 |
| Rework | 13 | 7 | 4 |
| Quality | Hidden |  | Surfaced |
| Suppliers | Disconnected |  | Involved |

# Cell Layout Types Airplane Game

**Key Areas**

Cell design, layout, job enrichment, productivity, line, cell, cross-functional working

**Manufacturing or Service?**

Manufacturing (and possibly Service)

**Overview**

This game or exercise is designed to stimulate discussion about different working arrangements. Assembly line working or cells often come to mind first – but there are other possible changes.

First, players need to time themselves doing each operation involved in making the airplane. Variation will be noticed.

Boring, repetitive, short-cycle work is less and less acceptable. But what are the alternatives? At one extreme a player can make the entire airplane. At the other extreme is the short cycle assembly line. There are options in-between.

This game or exercise is a good partner with the other airplane games. Once players are familiar with making the paper airplane, they can try out and discuss different layout or job tasking formats.

Players need to be familiar with making the paper airplanes. Once familiar, they make airplanes using five different layouts, each in a 12-minute period. Players collect the data, and then discuss.

**Summary of Learning Points**

- In building a complete airplane individually, there is likely to be big variation between players. Some players will do better than the assembly line average. Problems are inventory stocking points, lack of problem or progress visibility, and the cost of training. Variation is the big doubt.
- The assembly line: this is traditionally the most efficient. But is it? The problem is losses that occur due to imperfect balance. Alternatively WIP can build up, without kanban or CONWIP.
- Parallel: may be a fair compromise from traditional assembly line. Cost of equipment, training, inventory works against this type being adopted for longer, more complex lines such as cars. May be attractive for less complex, shorter lines.
- Transverse: an attractive compromise for job enrichment. Individual variation and traceability may be issues. Good for assembly. Less good where expensive equipment is required.
- CONWIP: an internal push system, hence this is more robust to timing variation, but does not allow inventory to build up. In effect, it is a multi-stage kanban, with the pull signal going directly from end to beginning.

**Players**

8 or 16 or 24 is best. In between these numbers some task balancing will be required.

**Approximate Time**

1.5 hours, including discussion

**Description**

Five types of layout are evaluated. These are shown on a game sheet.

1. Assembly Line: with 8 stations, maximum. The players should first balance (equalise) the work between themselves.
2. Individual working: each individual makes a complete airplane from beginning to end.
3. Two parallel assembly lines, with each line having half the stations but double the station time of the assembly line. The two lines can be made to pace one another.
4. Two transverse groups: in each group the individuals make half the airplane. The first group places half-completed planes into an intermediate WIP area from which the second group draws airplanes as and when ready.
5. CONWIP: this uses the Assembly line sequence, but simply lets one piece in as one piece is completed. Hence Constant Work In Progress (CONWIP). In between the stages there is push – move the work ahead as soon as finished – but overall there is pull – letting one piece in as one piece is let out.

    Note: A sixth type – bucket brigade balancing – is shown in a separate game.

**Player Instructions**

- Use the airplane assembly sheets. Make sure that each player is familiar with making a complete airplane satisfactorily.
- First, time five cycles of each operation of the 8 operation stages. Use (say) five different people to do the timing. This is a good exercise in itself. You will notice variation between the timers themselves. Take a reasonable average time – not the shortest or longest.
- Five types of layout will be examined – assembly line, individual working, parallel, transverse, and CONWIP. Your instructor will explain the difference. See the Figure.
- Take each type in turn and allocate the available players to operations for each type.
- If you have already done the 'Bucket Brigade' line balancing you could use this method to balance the 'Line', according to 'Parallel' and 'Transverse' types.
- Start with the following initial inventory for each type. This is an attempt make more valid comparisons between the various types.
    - Line: one partly completed airplane for each of the first four operations.
    - Individual: one sheet of unfolded A4 for each player
    - Parallel: one partly completed airplane for Operations 1 and 2, and 3 and 4, for each of the two lines.
    - Transverse: operations 1 to 3 completed for each of the four initial workstations.
    - CONWP: as for Line.
- Run each of the five types in turn for 12 minutes, and count how many good airplanes are made in each type. Also count the number of defectives.

The Lean Games and Simulations Book

- See the blank table. Discuss the five types under the headings given in the table. Complete the table. You may like to do this around a flipchart.
- Discuss the implications for your own business.

## Forms, Graphs, Tables, Typical Results

- Use the standard airplane operation sheets.
- Use the Cell Layout Types Airplane Game Sheets.

## Instructor notes and discussion

- Supervise the running of the game, but let the group members work out the workstation arrangements. This is a good part of the exercise.
- See the table for the consideration types. Facilitate completion of the table. See a few comments in the Summary of Learning Points above. There are many more that would be relevant to any particular situation.

## Equipment

- A ream of A4 sheets
- Sufficient red and blue felt pens – one for each player (for individual working)
- Sufficient rulers for quality control – one for each player
- 8 tables and chairs
- Flipchart with pens

## References and Further Reading

- Bicheno and Holweg, *The Lean Toolbox*, 4th edition PICSIE, 2009, Chapter 8
- Micheal Baudin, *Lean Assembly*, Productivity, 2002

# Bucket Brigade Line Balance Game

**Key Areas**

Line Balance, Cells, Productivity, Job Variety, Job Timing

**Manufacturing or Service?**

Manufacturing. Some potential in service operations such as clerical, hospital wards.

**Overview**

This is another game in the paper airplane series, using the standard airplane.

This game or exercise is designed to stimulate discussion about an alternative to traditional line balancing, often using timing or video. Boring, repetitive, short-cycle work is less and less acceptable. Likewise suspicion of externally timed operations, and errors (both deliberate and non-deliberate) bedevil line balancing. Here is an alternative.

The method also improves productivity (throughput) and leads to lower levels of inventory.

The method helps to overcome some of the other drawbacks of conventional line balancing, such as operators regarding their workstation as fixed, and the problem of a bottleneck stage that can effectively hold up an entire cell if kanban is used.

Of prime, and increasing importance, especially for Lean, is that mixed model line balancing is very difficult and / or inefficient. Mixed model lines – where several product variants in a family are made – is increasingly used in Lean manufacturing.

Bucket Brigade line balancing is a heuristic method that self-corrects for a variety of factors – such as job work content, operator variability (piece to piece, operator to operator, day to day) and demand.

This game or exercise is a good partner with the other airplane games. The game is ideally played before the Cell Layout Types exercise.

The game also gives practice in operation timing and in conventional line balancing.

**Summary of Learning Points**

- Bucket Brigade line balancing requires NO timing, or external analyst.
- It balances automatically, coping with changes in work content, operator variability and demand.
- Bucket Brigade automatically limits accumulation of WIP as well as increasing productivity.
- A job training priority sequence is suggested.
- Bucket Brigade balancing copes easily with changes such as operators going off to the toilet, and with job rotation.
- The method is often more efficient than traditional line balancing because time is used better by, for instance, not requiring exactly the same work activities in each work cycle.

- There is none of the working faster or working slower deceptions found with traditional timed work balancing. Rather, operators sort themselves out.
- Group dynamics tends to sort out slower or faster workers, but if not it does not matter.
- Good traditional line balancing should be done for several rates. With Bucket Brigade, this is automatic.
- Motivation should be better.
- Possible downsides are monitoring difficulties, and traceability.

## Players

Minimum 3, maximum about 6. Several onlookers can take turns, bringing the group total to perhaps 15.

## Approximate Time

1 hour

## Description

This is a demonstration exercise rather than a game. To start, the various operations are allocated amongst three operators using the heuristic method described. Productivity is measured. Then variations are added – an extra operator (total 4), one less operator (total 2), extra work content.

In each case, after a number of cycles, the work allocation settles down to a fairly stable state - but certainly not exactly the same for each product.

## Player Instructions

- The game requires very little preparation to get going, but each of the four operators (players) should be familiar with making the paper airplane. Each operator should make two complete airplanes, using the Operation Sheets, before the exercise proper begins.
- The Airplane Operation Sheets will be laid out in a sequence along a table, with a ream of A4 sheets at the beginning. Allow 1 metre between stations for clarity.

The Game is Played over two Rounds: Conventional and Bucket Brigade

### Round 1: Conventional Line Balance

- You will need timers for each observer. You can use stopwatches as found on iPad or iPhone apps. These are free apps. If you have stopwatches or timers, so much the better.
- Begin by timing each of the 7 operations (ignore the 8th inspection operation). To do this, let each operator do each operation five times. Player 1 does operation 1 and puts the semi-completed plane in a pile, then player 2, and so on. Time each individual operation. Fill in times on a chart such as shown on the Bucket Brigade Game Sheet. Then all players move to operation 2 and are timed. Then to operation 3, and so on.
- Quite wide variation between player times on each operation will be noticed. This is one of the problems with conventional line balancing.
- Balance the operation times between 3 players. Here, judgment is required. Typically, the line can be balanced by each worker doing about 45 seconds of work, as follows:

- o   Player 1 does operation 1 and 2 (station 1)
- o   Player 2 does operations 3 and 4 (station 2)
- o   Player 3 does operations 5, 6 and 7 (station 3)
- Notice the imbalance between stations!
- Then, run the round for 12 minutes with the players allocated between the operations. It is a push system, with planes accumulating between stations. Player 4 records defects.
- Record WIP inventory after 12 minutes, throughput, and defects.

**Round 2: Bucket Brigade**

- YOU WILL WANT TO WORK THROUGH THE FOLLOWING SEQUENCE SLOWLY, but players and observers will quickly catch on and begin to work at normal speed.
- Three players line up in front of the first three Airplane Operation sheets.
- Player 1 starts work at Operation 1 (Fold 1). The player passes the partly folded sheet to Player 2 who carries out Operation 2 and passes the airplane to Player 3.
- Player 3 completes Operation 3, then moves downstream to Operation 4, then 5, then 6, and so on until the last Operation is complete and the finished airplane is deposited in the Finished Product area. Player 3 then walks upstream – not to the beginning of the line, but upstream  - passing Operations 7, 6, 5, etc until Player 3 meets Player 2
- When Player 1 hands the part-folded airplane to Player 2, Player 1 returns to the beginning of the Line and starts making a new airplane.
- Player 1 hands Player 2 the second part-folded airplane. Player 2 does Operation 2 but then proceeds downstream to Operations 3, 4, etc., until Player 2 meets Player 3 walking back upstream.
- Player 2 then hands the part-completed airplane to player 3.  Player 2 then turns around and walks upstream towards Player 1. So, Player 2 does not complete making the full airplane.
- Player 3 takes the airplane from Player 2, turns around, and moves downstream completing all Operations until the airplane is complete. Then, once again, Player 3 walks upstream until he/she meets Player 2 moving downstream.
- When Player 1 handed the second airplane to Player 2, Player 1 once again returns to the start (Operation 1) and begins the third plane. This time, however, Player 2 is likely to be busy further downstream. So, Player 1 then works downstream doing Operations 2, 3 etc until Player 1 meets Player 2 walking upstream. Player 1 hands the part-made airplane to Player 2. Player 2 turns around proceeds downstream. Player 1 walks back to the beginning of the line and starts making the fourth airplane.
- Now the pattern of work is becoming established. Player 1 works downstream until meeting Player 2. Player 2 works downstream until he meets Player 3. When meetings (or airplane completion in the case of Player 3) take place, Players turn around and walk upstream, until they meet the earlier Player working downstream, or in the case of Player 1, when he reaches the beginning of the line.
- Eventually balance occurs.
- DO NOT 'LEAPFROG' other players. Work in your segment, even though the segment may vary somewhat for each plane.
- IF CATCHUP OCCURS, where a faster earlier player catches up with a slower later player, turn around and walk back upstream, either to the beginning of the line (for Player 1), or to where Player 1 is working (for Player 2).
- COMMUNICATE! This is a very important point on flexibility of working. The players should talk to one another about work progress. This includes split operations – see below.

**Further notes on Balancing**

- The Operation Sheets themselves contain several sub-operations – such as two folds. Generally, the whole Operation needs to be completed before handing on to the next Player. A Player should never take over a partly completed sub-operation – like half a fold.
- This may sometimes mean that there is waiting for a previous Player to complete an Operation. There are a number of possibilities:
  - Usually, just accept the delay.
  - Split an Operation into two or more sub-operations. For instance, Operation 3 can be split into two steps – fold along centre fold, and fold the wings over the body.
  - A third possibility is to have a kanban square around the usual point at which Players meet. The rule is, that if the square is occupied when the player reaches the square, the player must then wait for the next player to remove it before the first player begins walking upstream. This is not recommended because it works against the flexibility of the method.

**Forms**

- Use the standard Airplane Operation sheets.
- Use the Bucket Brigade Line Balancing Game sheet.
- Typical average times are as follows (but these vary wildly!)
  - Operations 1 and 2: 20 seconds each
  - Operation 3: 30 seconds
  - Operation 4: 15 seconds
  - Operation 5: 20  seconds
  - Operation 6: 18 seconds
  - Operation 7: 17 seconds

**Typical Results for 3 players in a 12 minute game**

- 'Balanced Line':  12 planes completed
- Bucket Brigade Line: 16 planes completed (a 33% productivity increase)
  - (Plus, of course, increased flexibility)

**Instructor notes and discussion**

- Layout a line with 1 metre between Airplane Operations sheets. These are the workstations.
- Have a few completed airplanes on display, and make sure that all players know how to make an airplane.
- Explain the motivation for Bucket Brigade Line Balancing. See the Overview and Learning Points sections above.
- Start with three Players. Explain the method, slowly.
- In Round 2, let the Line Balance develop. After a short while, when the game has settled down, measure the number of planes made in 6 minutes. Compare the results with Round 1.
- Then, add another player. Let it settle down, and again measure for 6 minutes.
- Then take two players out, so only two remain. Again settle down and measure.
- Restore to three, but alter the work. Take the decal marking out or have decals on every second airplane.
- Ask the group about training. New operators do not have to learn all the tasks. They can for instance start at the beginning or end gradually covering all the tasks. The experienced operators can cover during this learning period.
- Ask about the applicability of the method in their own organisation.

Discuss the advantages and conditions of Bucket Brigade Balancing. These are:

- Bucket Brigade is more productive, with lower inventory. Quality?
- Why? Because much time is wasted through an imperfectly balanced line and variation between workers and tasks.
- Bucket brigade is also more flexible – automatically coping with operators taking breaks, and different products.
- Standard work is important to remain consistent, both quality and productivity.
- Operators and Unions need to support cross-functional working.
- The work must have many possible hand-off points. Clearly, if there are long tasks that cannot be interrupted, the method is much less suitable. However, it may be possible to use the method before and after such operations.
- Operators and tasks need to be such that movement is possible. Sitting down tasks are a problem. Standing and walking is required.

## Equipment

Ream of A4 sheets
Red and Blue felt pens
Stapler
Timers

## References and Further Reading

Bartholdi and Eisenstein, 'A Production Line that Balances Itself', *Operations Research*, v44, n1, 22-34, 1996

# TWI Job Breakdown Exercise

## Key Areas

Job breakdown, Standard Work, SOP, Training within Industry, Job Instruction

## Manufacturing or Service?

All manufacturing. Many service operations.

## Overview

Using the standard airplane exercise, players construct a Job Breakdown Sheet, as used in Training Within Industry (TWI) job instruction. Job Breakdown is the first stage in Job Instruction (JI). JI helps create stability – an important reason why so many Lean interventions suffer sustainability problems.

This exercise is best done after one of the other exercises in this Airplane section – when players are familiar with making the airplane.

## Summary of Learning Points

- Learn the layout and use of a Job Breakdown Chart
- Understand 'Important Steps', 'Key Points', and 'Reasons for Key Points'
- A Job Breakdown Sheet is not an SOP

## Players

Any number up to (say) 15 can be facilitated

## Approximate Time

1 hour

## Description

Players go through the Airplane Operations sheets, and their own knowledge of making an airplane, and complete the Job Breakdown Sheet. The instructor will advise and facilitate.

**Player Instructions**

- Use The Airplane Operations Sheets, and your knowledge of making the paper airplane, to construct a Job Breakdown sheet that can be used to instruct a new worker on how to do a job.
- Identify the Important Steps that advance the work. These are the main logical steps. Note that not every step needs to be documented. For instance, you would not have a step called 'pick up paper' because that is obvious, and the 'how' is not critical to job quality.
- Then identify the 'Key Points'. These are points that can make or break the job, injure the worker (not applicable here), or make the work easier. For example in making an airplane it is very important to fold the A4 sheet exactly in half. A 'knack' point example is forming a fold between thumb and forefinger.
- Then fill in the reasons for the key points.
- It is an excellent exercise if the group can then use the job breakdown chart to instruct someone who has not previously taken part in one of these Airplane Exercises.
- The Steps, as specified by TWI, are:
    o Prepare the Worker. Explain the importance of the work, put the person at ease and in the right position. Encourage questions.
    o Present the Operation. Tell, show and illustrate one key step at a time. Do this again, stressing Key Points.
    o Try Out. Have the person do the job. Correct any errors. Then have the person do the job again explaining each important step. Then yet again explain each key point.
    o Let the person do the job, but check frequently and inform the person who to contact in case of problems.

**Forms and Typical Results**

A template is to be found in the Figures.

**Instructor notes and discussion**

It is recommended that the instructor read the first two parts of the reference before running this exercise.

**Equipment**

Ream of A4 sheets

**References and Further Reading**

Patrick Graup and Robert Wrona, *The TWI Workbook,* Productivity Press, 2006

# The Squares Game Set

The classic 'square game' and variations are used here for three relevant Lean concepts:

- Systems Thinking
- 5S
- Changeover Reduction

There are three games.

The origin of this manual, paper based, Squares Game for teams is unknown. The author first played it at Witwatersrand University, South Africa with David Brain in 1975.

# Squares Games 1: Systems Thinking

**Key Areas**

Systems, co-operation, implementation, teamwork, resistance to change, communication, parts and wholes. Holistic problem solving - right brain, as opposed to linear left brain.

**Manufacturing or Service?**

Both

**Overview**

This game is played by 5 players each of whom gets an envelope containing shapes. The objective is for each player to end with a square of equal size to all the others. The game is played in silence.

**Summary of Learning Points**

A 'total system' message. Easy to complete a few squares but getting all five requires participation by all. It may require players to break up their existing squares – some find this hard to do. Frustration at seeing what is obvious to some but not to others. 'Optimising' or completing one square whilst ignoring the others may be counterproductive.

**Players**

Groups of 5. More fun if there are 10, 15, 20, etc. players.

**Approximate Time**

30 minutes – more if good discussion.

**Description**

The game is played sitting down around tables, five players per table. The objective is for each player to make a square of equal size to the other squares. There are several ways in which squares can be made but only one way for all five squares to be made.

The game is played in silence. Players must not attempt to build another player's square, or to make more than one square. Players may not ask another player, in any way, for a piece. However, they may give up a piece by handing it to another player. Players remove pieces from their envelopes and attempt to build squares.

## Instructions

1. Each player is handed an envelope containing cardboard or paper pieces.
2. Ask the players to check that there are no remaining pieces in the envelopes.
3. The rules are explained. Namely:
   * Silence.
   * Players must not attempt to build another player's square, or to make more than one square.
   * Players may not ask another player, in any way, for a piece. However, they may give up a piece by handing it to another player.
   * No signaling.
   * A player may not help him/herself to another player's pieces.
   * There is unlimited time.
4. The instructor should see that the rules are maintained. The instructor may have to remind players of the rules.
5. Eventually, all five squares are completed – sometimes after 15 minutes or so.
6. Where there is more than one game in progress, the completed players should stand around (behind) uncompleted games, but maintain silence. This adds to the pressure on uncompleted games.
7. Discuss.

## Forms, Graphs, Tables, Typical Results

Use standard forms. See below.

Teams generally work out the solution, although some sometimes with delay. Let them struggle. Occasionally a team can simply not complete their squares, and gives up. In this case, give help by turning around one piece – usually one of the two 'Y' pieces, labeled B and E.

## Instructor notes and discussion

After all teams have completed their squares, ask for comment. Write these down on a flip chart. You may have to lead, by asking 'what did you feel about…' Typical comments include:

* Realising that the whole team needs to participate. (A systems message.)
* Realising that completed squares may have to be broken up, in order to complete all five. (A systems message, often occurring in Lean implementation.)
* Frustration at seeing the solution, but having to wait patiently. (Often occurs in Lean implementation. Self discovery is much better than being told.)
* Frustration at not being able to communicate.
* Did a player withdraw from the group after finishing his/her square? (Sometimes occurs in Lean implementation.)
* Feeling pressure to complete. Sometimes there is pressure on 'uncompleted' players, but the pressure should be on 'completed' players to break up their squares.
* Giving all pieces to one player. Sometimes done in frustration. 'OK, you sort it out!'
* Resentment of watching players. (How do you feel when some 'smart ass' is looking over your shoulder?)

Then ask players to recount their own implementation experiences of implementation. Suboptimisation? Breaking up an apparently perfectly good sub-system? Did everyone participate? Was there a blocker? Was there frustration? How did you feel?

Deming spoke of four reasons why many implementations fail. The four are:
- Lack of appreciation of system interaction
- Variation
- Scientific problem solving
- Psychology

This game illustrates the first and last of these. The game also illustrates holistic problem solving. This type of problem solving is essentially different from the linear type as embodied in PDCA. Both types are needed. There is sometimes much attention given to the PDCA / DMAIC type, but insufficient to holistic 'right brain' thinking. This may be worth quite a bit of discussion.

(Variation is illustrated in the Measures and Targets Game.)


**Equipment**

Front and back of five completed squares are given. These should be cut out, labeled on front and back, the pieces separated and inserted into envelopes A to E. The pieces are labeled according to which envelopes they are to be inserted into.

Five A5 size envelopes, labeled A to E, each containing appropriate pieces.

Flip chart and pens.

Cut out front and back of each square, and either glue the pages together or use cardboard or thick paper. In the latter case, please label front and back with the letters. If this is not done, it gives a strong clue to the solution.

Note: do not include the extra pages, labeled 'extras', for the Systems game. These are used for the 5S and Changeover games.

**References and Further Reading**

For discussion on two types of problem solving, see Karl Albrecht, *Practical Intelligence*, Jossey Bass, 2008, Chapter 12. You may like to compare this with the A3 type discussed earlier.

Many references on Systems, including by Seddon, Deming, Checkland, etc.

For an amusing review of systems 'laws', part serious part fun, see John Gall, *The Systems Bible*, General Systemantics Press, 2006

# Squares Games 2: 5S

## Key Areas

5S, Waste, Housekeeping, Productivity, Variation reduction, Standards

## Manufacturing or Service?

Both. 5S is applicable in all direct manufacturing, but not all service and administration.

## Summary of Learning Points

- When things are in a mess productivity is affected and variation is greater
- How much time is lost looking for things?
- The 5S's are Sort, Simplify, Scan, Standardise, Sustain. There are variations on this; for example, Simplify may be replaced by Straighten; and Scan by Shine.

## Players

Any number, in teams of 3 to 5 players each.

## Approximate Time

1 hour

## Description

The teams must assemble 5 squares from pieces contained in 5 envelopes. The first round is typically wasteful and time consuming. Each team then works through the 5S approach, putting forward suggestions as to how the contents of each of the five envelopes should be stored. A second round follows, again using 5 envelopes but with non-essential pieces eliminated and markings added.

## Instructor notes and discussion

Unlike the Systems Squares game, each team is simply given 5 envelopes (labeled A through E) and told to make 5 squares of equal size. Talking is allowed. Use the pieces from the 7 squares (5 Squares and 2 Extras).

After the squares have been assembled, explain the concept of 5S. Sort, Simplify, Scan, Standardise, Sustain.

Of these, the teams can certainly Sort, Simplify, Standardise, and think of ways to Sustain by, for example, an audit process or competition.

Explain that, for storage reasons, the pieces must continue to be kept in 5 separate envelopes. The pieces must be kept separate. However, contents of envelopes may be changed, and pieces may be marked.

The following steps need only brief explanation.

- Sort: identify and throw out unneeded pieces
- Simplify: sort the remaining pieces? Other ideas?
- Standardise: a standard procedure to store, put away and use? This should be written up as a SOP (standard operating procedure)
- Sustain: ideas to audit and keep in place?

Give the team a specific time to complete the work. 30 minutes is more than adequate.

They will normally eliminate extra pieces and mark up the essential pieces.

Each team should then report back to the others.

**Equipment**

Front and back of five completed squares are given. These should be cut out, the pieces separated and inserted into envelopes A to E. The pieces are labeled according to which envelopes they are to be inserted into.

For this game use ALL SEVEN squares. The additional two squares contain extra pieces that are not required to build the squares, but are there to delay and confuse the teams. Of course, these pieces will need to be identified and eliminated.

Cut out the pieces from the additional two squares and insert in the appropriate envelopes A to E.

Flip chart and pens. One for each team preferable.

Cut out front and back of each square, and either glue the pages together or use cardboard or thick paper. In the latter case, please label front and back with the letters. If you don't do this, it gives a strong clue to the solution.

**References and Further Reading**

- John Bicheno and Matthias Holweg, *The Lean Toolbox, 4*[th] *edition, The Essential Guide to Lean Transformation*, PICSIE Books, 2009, pages 78-82
- A detailed discussion of the role of 5S in Lean is in Gwendolyn Galsworth, *Visual Workplace, Visual Thinking*, Visual Lean Enterprise Press, 2005

# Squares Games 3: Changeover Reduction

## Key Areas

Changeover, setup, SMED, waste, batch sizing

## Manufacturing or Service?

A key concept in Lean Manufacturing. May be relevant in some types of service – for instance preparing an operating theatre, or making up a room in a hotel.

## Overview

Changeover reduction is a key enabler for smaller batches in Lean manufacturing. It may also save capacity. This game uses an extended version of the Squares Game to simulate a changeover. The changeover team is presented with 5 envelopes representing the components (dies, tools, etc) that are required, and a 'heat element' required to 'pre-heat' the die. The time taken is measured, and activities are mapped. A briefing on 'SMED' is given. Waste and time is then reduced. General lessons are given.

## Summary of Learning Points

- The Shingo SMED sequence: separate external and internal activities, shift as many internal activities to external, remove wastes, reduce internal times, reduce external times.
- Mapping and analysis of changeover activities.

## Players

4 to 6 per team.  Can have multiple teams. Good to have several teams and have them compete.

## Approximate Time

40 to 90 minutes, including discussion. Depends on extent of mapping and analysis activity, and presentations back to the other teams.

## Description

- Divide the group into teams of 4 to 6.
- Each team will need a table where the 'changeover' will take place.
- Explain to the teams that, when the game begins, a batch has just been completed and the dies for the next batch must now be prepared.
- Prepare the five envelopes, each with pieces, and the Heat Treat sheet. These make up a set. Have a set for each team, clearly marked with the Team number. Arrange the envelopes in different corners of the room. All the 'A' envelopes together, 'B' envelopes together, and so on. Note: do not put all the envelopes belonging to one team together. They must be separated, to simulate going to fetch tools and dies from different stores.
- Explain that each team is to assemble 5 equal sized squares, as quickly as possible, arranged alongside each other, with the circle (heat treat) square in the middle. That is the third square from either left or right. Each team must time itself.

- Thereafter, briefly explain the SMED process. (See The Lean Toolbox, 4th edition.)
- Now each team must place ALL pieces (including extras) back in the envelopes, and the envelopes back to the original locations.
- Begin the changeover again, in own time. Analyse and time each element. Draw a 'map' or activity sequence. Use 'post-its'. Draw a spaghetti diagram of movement. Remove unneeded pieces. The contents of the envelopes may change, but the locations of the five envelopes must remain. This constitutes moving internal to external times, and waste removal. Mark up the pieces. This constitutes reducing internal time. Have each team practice the 'changeover'. Practice is part of a slick changeover.
- Note that the Heat Treatment stage includes adjustment – which is often a time consuming activity in a real changeover.
- Have each team document the standard procedure.
- Have each team demonstrate their new procedure.
- Award a prize to the winners?
- Discuss application in own business.

**Player Instructions**

- A previous batch of product has just been completed. Your task is now to prepare for the next changeover. The changeover will be considered complete when six equal sized squares are assembled in a rectangle on your table. The top right square must contain the square with the circle, duly 'heat treated'.
- Note the Heat Treat requirements given on the Heater Sheet
- Note the instructions as given on the 'Changeover Instructions' sheet.
- You will find five envelopes at five locations around the room. These constitute the die and tool stores. The Heater card is on your table.
- Your team must time and record the time to assemble the six squares, including 'heat treat'.
- Once you have timed the initial changeover, the instructor will say a few words on the changeover process.
- Put ALL pieces back in the original envelopes, with the original piece locations. That is pieces 'A' in the A envelope, etc. Return the envelopes to their original locations.
- Repeat the changeover. But now, document and time the steps. Draw a process map. Use post-its. Draw a spaghetti diagram of movement. Remove unneeded pieces. The contents of the envelopes may change, but the locations of the five envelopes must remain. This constitutes moving internal to external times, and waste removal. Mark up the pieces. This constitutes reducing internal time. Have each team practice the 'changeover'. (Practice is part of a slick changeover.)
- Document the procedure. Write up the SOP.
- Be prepared to present your solution.

**Forms, Graphs, Tables, Typical Results**

A dramatic reduction from around 10 minutes to around 1 minute or less, is usually achieved.

## Instructor notes and discussion

- Take the opportunity to ask the teams about the potential for reduction at their workplaces.
- Remind players about internal and external times, and about identifying waste.
- Technology can be important in changeover – like handling the circle, and pre-heat, but moving internal activities to external is where the greatest opportunities generally lie.
- Adjustment is important in changeover. 'One touch' adjustment is desirable.
- Consistency. It is not only the speed of the changeover, but also how consistent it is.
- Further opportunities always exist. The envelopes could be moved to point of use.
- Documentation is essential. Mapping is useful.

## Equipment

- You will need a fair size seminar room. A lecture room with fixed seats is not ideal.
- Five envelopes for each team, containing the six squares. Use all Squares from the Squares Game, and the additional Square for the Changeover Game.
- Cut out and place in five marked envelopes. Note, the special changeover square with the centre circle cut out, must be used.
- A 'Heater' sheet must be given to each team
- A flip chart for each team.
- A timer or stopwatch for each team. You may be able to use timers built into mobile phones or iPhone.
- A pad of post-its for each team.

## References and Further Reading

Bicheno and Holweg, *The Lean Toolbox*, 4th edition PICSIE, 2009, pages 89-92

R.I. Macintosh et al, *Improving Changeover Performance*, B-H, 2001

# Introduction to Dice Games

The Dice Games are a set of games, each of about 30 minutes, that form a form a backbone of understanding of Lean in general and Lean scheduling in particular.

A selection can be played to bring out appropriate points in relation to progress on your Lean journey. There is a lot to learn from them!

Each game requires a minimum of 6 players, but in many cases several games can be run in parallel, so that perhaps 30 players can participate if there are a number of facilitators. Basically, only one or two dice per player is required together with a movable part such as Lego bricks or even fasteners such as washers, nuts or bolts, or even wrapped sweets that be eaten as a reward at the end of the session!

Dice Game 1 (Muda, Muri, Mura) could well be the most important game in the book for your understanding of Lean.

The original dice game was one developed by Creative Output in the mid 1980s to illustrate various Theory of Constraints or 'OPT' Principles. A modified version of this game is Game number 2.

The original TOC dice game (described in Dice Game 2 and those following) is essentially a 'batch and queue' game in as far as variable batches are made and moved forward to the next station.

For this edition, a new set (beginning with Dice Game 12) has been developed. These games are developed around one piece flow. As soon as a single part (or part) is completed it is moved forward and is available for production at the next station. This also allows some of the very powerful Factory Physics concepts to be explored using games.

**References and Further Reading for all Dice Games**

John Bicheno and Matthias Holweg, *The Lean Toolbox*, 4th edition, *The Essential Guide to Lean Transformation*, PICSIE Books, 2009, Chapter 9 Lean Scheduling and Chapter 10, Theory of Constraints and Factory Physics. Note: A New Edition will be published in 2015.

Wallace Hopp and Mark Spearman, *Factory Physics*, 3rd Edition, McGraw Hill, 2008

Wallace Hopp, *Supply Chain Science*, Waveland, 2008

John Darlington, *The Dice Game*, MSc Lean Operations Notes, Buckingham University BLEU, 2008

John Arthur Ricketts, *Reaching the Goal*, IBM Press, 2008

# Dice Game 1: Muda, Muri, Mura

**Key Areas**

Waste, Variation, Capacity Utilisation, Queuing, Lead Time

**Manufacturing or Service?**

Both. For manufacturing this would be a typical make to order environment. For service, any queuing situation such as in a bank or call centre. The lessons are, however, fundamental to ALL types of manufacturing and service. Queuing is pervasive in operations, so understanding the causes and interactions is of fundamental importance.

This game illustrates 'Kingman's Equation' that has been called 'The Equation of Lean'. The relationships and curves that are developed are discussed in several publications:
- Hopp and Spearman, *Factory Physics*
- Wallace Hopp, *Supply Chain Science*
- Donald Reinertsen, *The Principles of Product Development Flow*
- Niklas Modig and Par Ahlstrom, *This is Lean*

**Overview**

The game is played by pairs of players. One player plays the part of customers (or orders), the other plays the process serving the customers or processing the orders. Each pair generates a set of data that together generate the classic exponential curve of average queue (or lead time) against capacity utilisation. From this curve, fundamental relationships between lead time and variation (mura or amplification) and utilisation (muri or overload) can be discussed. The curve is also referred to by Kevin Duggan as 'the chaos curve'.

**Summary of Learning Points**

- Variation is the enemy
- Reducing utilisation (say by cutting failure demand or diverting work) is very effective, particularly at high levels of utilisation.
- Overload begins at less than 100% utilisation
- The greater the variation, the longer the lead time
- At low capacity utilisation variation is not so important
- At high capacity utilisation, there is greater uncertainty about lead time
- Uncertainty about lead time is at least as big an issue as lead time itself.

**Players**

A minimum of 8 players, in 4 pairs. Best is at least 6 pairs with 12 players.

**Approximate Time**

30 minutes, but discussion can go on much longer because of the fundamental lessons.

**Description**

Players are asked to form themselves into pairs. Give each player one dice. Each pair of players plays their own independent game.

One player plays the role of all the customers, generating all demands or orders. The other is the process, responding to the demands. In manufacturing, the process could be an entire plant or a single manufacturing stage. In service, the process is any situation where there are queues – health, banking, shops, call centres, repair and maintenance.

Introduce the game by asking if anyone has a situation with
- Absolutely constant, level demand?
- No variation in process time, including no breakdowns?

Of course, such situations are virtually unknown. That is why each player is given a dice.

This game assumes that no 'anticipation' finished goods inventory is possible. (So this game is for make or assemble to order). In service, of course, one cannot 'inventory' service – like serve a customer in advance. The assumption is that customers (or oders) that are not served immediately do not go away, but remain in the queue.

**Player Instructions**

The game is played over 4 rounds of 20 periods each. In real life, a period may be a day, an hour, or any interval. In each period, each player rolls their dice. The customer's dice represents the demand in that period. The process dice represents the capacity that is available that period to meet the customer's demand.

In round 1, the process player adds 3 to the dice roll number, in each period.
In round 2, the process player adds 2 to the dice roll number, in each period.
In round 3, the process player adds 1 to the dice roll number, in each period.
In round 4, the process player adds 0 to the dice roll number, in each period.

Each period, each pair of players should record the number (people or jobs) in the queue, waiting for service, at the end of the period.

If there is a person (or job) in the queue at the end of the period, those people or jobs are carried over to the next period.

**Example:**

- Round 1, period 1:
  Customer rolls 4, process rolls 2. Process adds 3. 2+3=5, hence 0 in queue
  0 carry over
- Round 1, period 2:
  Customer rolls 2, process rolls 5. Process adds 3. 5+3=8, hence 0 in queue
  0 carry over
- Round 1, period 3:
  Customer rolls 5, process rolls 1. Process adds 3. 1+3=4, hence 1 in queue

1 carry over
- Round 1, period 4

  Customer rolls 3, so demand is 3+1 =4; process rolls 3. Process adds 3. 3+3=6, hence 0 in queue
- Etc

Continue for 20 periods. Then start round 2, with zero in queue. Then rounds 3 and 4.

After playing 4 x 20 = 80 periods, each pair should add up the total number in the queue over all 20 periods. The average is calculated by dividing the sum by 20.

**Forms, Graphs, Tables, Typical Results**

Number in Queue at end of Period

|  | Round 1 | Round 2 | Round 3 | Round 4 |
|---|---|---|---|---|
|  | +3 | +2 | +1 | +0 |
| 1 | 0 | 0 | 0 | 1 |
| 2 | 0 | 0 | 2 | 0 |
| 3 | 1 | 0 | 1 | 2 |
| 4 | 0 | 0 | 0 | 0 |
| 5 | 0 | 1 | 0 | 3 |
| 6 | 0 | 0 | 1 | 2 |
| 7 | 0 | 1 | 1 | 3 |
| 8 | 0 | 0 | 0 | 6 |
| 9 | 0 | 0 | 1 | 4 |
| 10 | 0 | 0 | 2 | 4 |
| 11 | 2 | 0 | 0 | 8 |
| 12 | 0 | 0 | 0 | 5 |
| 13 | 0 | 2 | 3 | 11 |
| 14 | 0 | 1 | 1 | 11 |
| 15 | 0 | 0 | 0 | 9 |
| 16 | 0 | 0 | 2 | 13 |
| 17 | 0 | 0 | 0 | 12 |
| 18 | 0 | 0 | 0 | 10 |
| 19 | 0 | 0 | 0 | 8 |
| 20 | 0 | 0 | 1 | 11 |
| Sum | 3 | 5 | 15 | 123 |
| Ave Q | 0.15 | 0.25 | 0.75 | 6.15 |

## Instructor notes and discussion

Either before or during the game the instructor should draw the following axes on a flip chart.

Then, plot the results (average queue) on the graph/axes. Plot the 4 points from each pair. A typical set of results is shown:

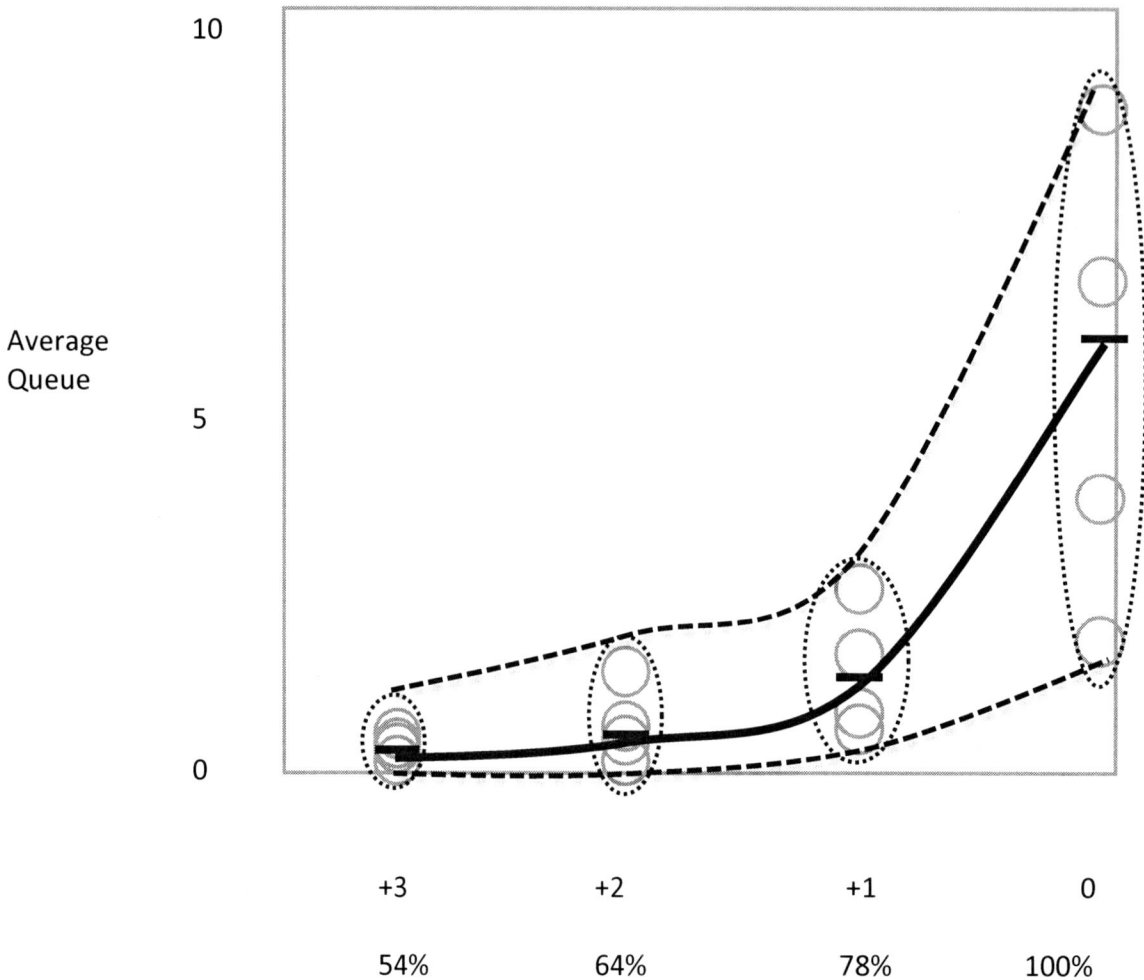

| | +3 | +2 | +1 | 0 |
|---|---|---|---|---|
| | 54% | 64% | 78% | 100% |

## Instructor-led Discussion

The instructor should either calculate or estimate the average of all points at each of the four capacity points +3, +2, +1, 0. Mark these average points on the graph. Mark off the range, as shown in the figure.

The instructor should then draw 3 lines on the graph: the upper line, passing through the highest point at each capacity point; the average line, passing though the average points; the lower line passing through the lowest point at each capacity point.

You will obtain a funnel-shape envelope – narrow at the +3 end and much wider at the 0 end. Do make a 'big deal' of this funnel. It shows increasing uncertainty. Uncertainty is damaging. You may

The Lean Games and Simulations Book

have a big queue or a little queue. It is a matter of chance.  You don't want to become known as an unreliable supplier: sometimes short leadtime, sometimes long.

The average line shows the average queue or lead time against utilisation. Utilisation is the average arrival rate (always 3.5 – the average dice roll) divided by the average capacity available (average dice roll plus 0, 1, 2 or 3). So +3 is (3.5/ (3.5+3) = 54%).

Why do queues grow? Because unused capacity can never be recovered. However, customers keep coming at the average rate and are never lost.

Notice that:

- Variation is the enemy. If there were no variation, the graph line would run along the base. There would be no queue. For instance, if everyone had a dice with the same one number (say 4), then with +3 there would be 3 arrivals each period, and a capacity of 7 each period, so zero queue.  This is MURA.
- But variation has little effect at low utilisation levels. As utilisation increases, the queue becomes less predictable.  At 100%, you may have a lucky run, or you may have a most unlucky run. You just can't tell! (Or, as Dirty Harry might say, if working at higher utilisation levels, 'Are you feeling lucky, punk?')
- The greater the variation the greater the queue or lead time and the greater the uncertainty.
- Hence reducing variation is most worthwhile – 5S, standard work, capability.
- The graph is exponential – it increases sharply against utilisation.  This is of foremost importance. If a machine or process is heavily loaded the lead time increases exponentially. This is MURI.
- Queues build up sharply at less than full utilisation. Or, overload (MURI) begins at less than 100% utilised. Therefore, pushing utilisation too high is counter productive!
- Rule of thumb: be very cautious about loading above about 85%. Manufacturing can get away with a bit higher, because of less variation. Service should be a lower than this.
- You need to provide extra capacity in manufacturing and service. If capacity is just enough to cope with average demand you will be in trouble!
- Major causes of waste (MUDA) are MURI and MURA!
- Kevin Duggan, an authority on Lean scheduling, calls this graph 'the chaos graph'. If too heavily loaded, everything starts to go wrong. Delivery dates are missed. Customers are kept waiting.

- Finally, Hopp and Spearman, from *Factory Physics*, propose the following two Laws. *"Increasing variability always degrades the performance of a system"* and *"If a station increases utilisation without making any other changes, average WIP and cycle time will increase in a highly nonlinear fashion."* Both are well illustrated in this game.

**Equipment**

- One dice per player. Players play in pairs.
- Each pair should keep a table, for four rounds and 20 periods as shown above.
- Ask players to record the outstanding queue length after every period.
- Have a flip chart for recording the queue vs. utilisation graph as shown in the example above. Draw the axes of the graph, with the average queue axis range from 0 to 10. (Sometimes you will get average queue lengths much more than 10.)
- To draw the graph, go around the pairs of players and ask for the average queue for +3. Plot these on the graph. Then repeat for +2, for +1, and for 0.

**A link with other Lean Concepts:**

1   Utilisation   = Load/Capacity

               = (Value Demand + Failure Demand  ((or Rework))/(Base Capacity – Waste)

So

- Decreasing Failure Demand or Rework decreases load, so decreases utilisation and (see graph) reduces queues or lead time (exponentially!)
- Reducing waste increases capacity, so decreases utilisation, and decreases lead time, exponentially.

2   If you only work on reducing waste and internal variation you are only attacking half the problem. The other half is external variation of demand.

For more explanation see *The Lean Toolbox* or *The Service System Toolbox.*

# Dice Game 2: Basic Batch Dice Game

## Key Areas

Capacity of a Line, 'Laws of Factory Physics', TOC Principles, Capacity

## Manufacturing or Service?

Mainly Manufacturing; some Service. Any sequential operations.

## Overview

A sequential line of workstations is simulated. The workstations all have variation. The game illustrates what Goldratt calls 'statistical fluctuation and dependent events'. As the game is played inventory (and lead time) builds up. Nominal (average) throughput is not achieved. As an aside, a warning is given about reward and punishment in meeting objectives.

## Summary of Learning Points

- Variation is the enemy
- Output is less than nominal capacity
- Deming's warning (natural variation)
- In an 'uncontrolled' system, inventory increases steadily
- Accumulation of inventory does not necessarily indicate a bottleneck
- The problem of 'balancing a line' ('Balance flow, not capacity')

## Players

6 minimum; up to 11. With 12 or more, start a second line.

## Approximate Time

40 minutes.

## Description

A sequential 'assembly line' is set up. Products move sequentially along the line. The game is played over 20 periods. Each player rolls a dice each period and passes products to the next player. The target of nominal capacity is never reached, and inventory increases as the game progresses.

## Instructions

Form a line of players at a long table. Straight is better, but L shape is OK. Give each player a dice and four pieces or products. The products may be Lego bricks, nuts, bolts, or (maybe) wrapped sweets. Imagine the game goes from left to right.

Each player has two squares before his or her workstation. The squares are labeled 'Receive' on the left and 'WIP' (Work in Process) on the right. The initial inventory is placed in the WIP squares.

The player at the beginning of the line does not have two squares, but instead has a bucket of pieces, to represent raw material.

Inventory (products / pieces) flows along the line from left to right.

Ask the players 'What is the average capacity per period?' Answer: it is 3.5 – or the average of dice rolls. This is why the initial inventory is 4 pieces.

The instructor should write a column of numbers (1 to 20) and two further blank columns (two for each team) on a flip chart. Each period, the instructor records the number of pieces that are completed (produced) by the last (rightmost) workstation. The instructor also writes the cumulative number of pieces produced on the chart.

Tell the players that they must all play simultaneously, as instructed by the instructor. The instructor will call out the period number.

The game steps are

- Each period the instructor will call out the period number.
- Each player rolls his dice. That is the available capacity that he has for that period.
- Each player draws on the inventory that he has in the WIP square. He passes on a quantity equal to the dice throw, if possible. If there is less inventory in the WIP square than the dice roll, then pass on all the inventory in the WIP square.
- Do not use any inventory from the Receive square. This is only available next period.
- 'Pass on' means place the inventory in the next player's downstream 'Receive' square.
- After all players have placed their inventory in the next player's Receive square, move the inventory from your own Receive square into your WIP square. There should then be no inventory in the Receive squares.
- If you do not have any inventory in your WIP square at this stage, you or a previous player has made a mistake.
- Wait for the next period and repeat.
- <u>For the first player</u> (at the left hand end), each period, roll the dice and place that quantity of pieces in the Receive square of the second player.
- <u>The last player</u> should inform the instructor of the number of pieces that are completed.
- Some piece recycling may be required. Finished items go back to the start.

Tell the players that nominal capacity of the line is 20 x 3.5 = 70 pieces. Ask the players to estimate the total output from the line. Write these estimates down on the flip chart.

<u>Halfway through the round</u>, after period 10, stop and review. Point out that the output thus far is less than the nominal 35.

Also record the inventory in the WIP squares of each player. Ask the players to notice that the 4 pieces each that the game started with has now become varied – some with less than 4, some with more.

Add up the total inventory in the WIP squares. Very likely, this total will exceed the initial quantity in the WIP squares. i.e. (total players – first player) x 4. Point out that inventory has increased.

Now do the 'Deming' charade. Go along the WIP squares. Find out who has the most inventory. Say 'That is not good enough. You will have to improve! I will be watching you!' Then find the person with least inventory in the WIP square. Say, 'Great performance. You are just the sort of employee we are looking for! Well done! Put it there!' (Shake his hand.)

Then, do the second 10 periods as for the first.

**Forms, Graphs, Tables, Typical Results**

Results after period 10

| Player | 2 | 3 | 4 | 5 | 6 | 7 | 8 |
|--------|----|---|---|----|---|---|---|
| WIP | 10 | 7 | 3 | 12 | 5 | 1 | 4 |

Total WIP = 42

Initial WIP = 7 x 4 = 28

Results after period 20

| Player | 2 | 3 | 4 | 5 | 6 | 7 | 8 |
|--------|----|---|---|---|---|---|---|
| WIP | 13 | 2 | 8 | 7 | 2 | 4 | 7 |

Total WIP = 44

Initial WIP = 7 x 4 = 28

| Period | No out | Cumul. |
|--------|--------|--------|
| 1 | 4 | 4 |
| 2 | 3 | 7 |
| 3 | 5 | 12 |
| 4 | 1 | 13 |
| 5 | 3 | 16 |
| 6 | 2 | 18 |
| 7 | 5 | 23 |
| 8 | 1 | 24 |
| 9 | 2 | 26 |
| 10 | 3 | 29 |
| 11 | 2 | 31 |
| 12 | 1 | 33 |
| 13 | 4 | 37 |
| 14 | 3 | 40 |
| 15 | 2 | 42 |
| 16 | 1 | 43 |
| 17 | 6 | 49 |
| 18 | 2 | 51 |
| 19 | 2 | 53 |
| 20 | 3 | 56 |
|  |  |  |

**Instructor notes and discussion**

Point out that

- The target of 70 has not been reached. It is never reached! Why? Because of variation, not only at each workstation, but also because of variation higher upstream. Goldratt calls this 'statistical fluctuation and dependent events. A general conclusion is that OUPUT < NOMINAL CAPACITY.
- Inventory has increased. It always does! Why? Because what is being let in is an average of 3.5 but the rest of the line together is restricting flow to below 3.5.
- Variation is the enemy! If there was no variation – say everyone had a dice with only a 3 (that is less than the average of a normal dice), output would have increased and inventory would have stayed the same.
- So, with variation, if you expect to reach your target, you will frequently be disappointed. Many managers don't realise this fundamental point!
- Accumulation of inventory does not necessarily indicate a bottleneck. It may be too soon to judge. Over the long term however, accumulation may well indicate a bottleneck.
- Deming. Often, as in the example, the player who was given a warning has improved. Clearly, then, management warnings and motivations work, don't they? So management should get a bonus! Clearly, this is rubbish. But how often does that happen in a company. Deming was fond of saying that many managers don't understand variation and are apt to hand out undeserved rewards and punishments. Of course, this is the difference between common cause and special cause variation. Here we have common cause or natural variation. Rewards and punishment is not appropriate. And, perhaps the player who was 'the best' is no longer so. Sometimes, as Deming said, praise has 'gone to his head…….'
- The problem of 'balancing a line': Goldratt said 'Balance flow, not capacity'. Here we have a perfectly balanced line, but it fails to produce the target output. Much effort is put into balancing real lines or cells. But, as illustrated here, this may be futile if there is much variation. A balanced line can only be effective if there is minimal variation. It is not that balanced lines don't work, it is that they don't work well with variation. There are better ways. See below.
- The longer you play the game, the more inventory there will be. Eventually, after a long period, there would be so much inventory in the system, and so much inventory in front of each work station, that the effects of variation will be buffered by inventory. The game will then reach equilibrium or 'steady state' with average input being matched by average output. In this case, then, we would have Little's Law. That is WIP = Throughput x Lead time.
- Once again, a *Law of Factory Physics* from Hopp and Spearman is "*Variability in a production system will be buffered by some combination of:*
    1. *Inventory*
    2. *Capacity*
    3. *Time*

So, if you have variability you will pay for it by inventory, capacity, or time. No way out of this!

## Advanced Note

If there was no variation (say, if all players consistently rolled a 3), output would be a predictable 3x20= 60. This can be achieved with exactly 3 pieces of inventory between stations. If inventory is less than 3, output of 60 will not be achieved. If it is more than 3, output will not increase but will remain at 60. This is what Hopp and Spearman, in Factory Physics, refer to as the 'best case'. But if there is variation, 60 will not be achieved unless there is a huge amount of inventory between stations. A graph of throughput vs inventory will show an asymptotic curve starting at the origin and gradually approaching the 'best case' line. This curve can be equated with what Hopp and Spearman refer to as the 'practical worst case'. Below this line is the 'Non Lean' zone. Between these two lines is the 'Lean' zone. Where are you?

Note: This concept is easier to explain and explore in the Single Piece Flow Dice Games, given later.

## Equipment

4 pieces (products) per player, plus 30 pieces minimum in the upstream raw material store. One dice per player. Flip Chart. The pieces (products) could be lego bricks, nuts (as in bolts), or wrapped sweets.

# Dice Game 3: Process Variation Reduction

### Key Areas

- This game follows directly from Dice Game 2
- Reducing variation improves throughput and decreases WIP inventory
- A big reason for standard work, 5S, Six Sigma

### Manufacturing or Service?

Important for both

### Overview

The game should be run immediately after Dice Game 2. The average capacity is maintained at 3.5 (the average dice roll) but variation is reduced by not allowing a 1 or 6. If a 1 or a 6 is thrown, the dice must be rolled again.

The game should not be run before players have played Dice Game 2.

Instead of running the game, the instructor may like to explain it. It is, however, far more convincing to run it and see.

Of course, the game can be re-run with even further reduction in variation – by not allowing 1, 2, 5, or 6 as legitimate rolls. This again will improve performance yet again.

As noted in Dice Game 2, if all players had a 'dice' with only a 3, then output would be exactly 60 units and WIP would remain unchanged at 4 per player. A point worth noting is that, in this case, the WIP can be reduced to 3 per station without affecting output.

### Summary of Learning Points

- Almost invariably, compared with Dice Game 2, throughput increases slightly and WIP falls.
- Reducing variation is NOT just about quality, but has a big impact on the necessary inventory levels and throughput.

### Players

Same as Dice Game 2

### Approximate Time

20 minutes, since players will understand game mechanics from Dice Game 2.

**Description, Instructions, Forms, Equipment**

As for Dice Game 2 but dice rolls 1 and 6 not allowed. If a 1 or 6 is rolled, the dice roll must be repeated until a roll between 2 and 5 is obtained.

No need to stop at period 10.

**Instructor notes and discussion**

- Ask the players to predict the outcome. Write these predictions on a flip chart.
- Ask why they have predicted these numbers.
- After running the game review the predictions.
- After running the game, make the point that many still think of 5S and standard work as tools for quality. They are, but the big payoff is often in inventory reduction and throughput.

# Dice Game 4: Reducing Input Variation

## Key Areas

A game that follows directly from Dice Game 3 and/or Dice Game 2

## Manufacturing or Service?

Important for both

## Overview

This game shows the effect of stabilising input. Process variation has received much attention but the arrival rate of work far less so. If the arrival rate is stabilised, the impact on inventory and lead time can be significant. Surprising perhaps?

A warning: This game does not necessarily produce dramatic results.

## Summary of Learning Points

- Pay attention to the stability of orders and/or the launch rate of work into production.
- Stability or launch of work can cut inventory (lead time) and (to a lesser extent) improve throughput.

## Players

6 minimum. With 12 or more, play two parallel games.

## Approximate Time

20 minutes, since players will understand game mechanics from Dice Game 2

## Description

The Dice Game 2, with limited variation by the first (upstream) player. Total inventory goes down, compared with Dice Game 2. Inventory is equivalent to lead time, so lead time improves.

## Instructions, Forms, Tables

Each player has one dice. Play Dice Game 2, except that the gateway (upstream, first) player may only roll a 3 or a 4. If any other number is rolled, the player must roll again. All other players roll their dice (1 to 6) as in Dice Game 2. The piece movement mechanics are exactly as in Game 2. All players again start with 4 pieces each.

All other instructions, forms and tables remain unchanged. Pause after period 10 and make a comparison with the results after period 10 in Dice Game 2.

**Instructor notes and discussion**

Explain the game mechanics, but do not otherwise tell the players what the game is about.

Ask the players what they expect to happen.

After playing, review the results.

Explain that stabilising the entry of pieces limits surge activity. Of course, the longer the line, the more that surges become established. At period 10, there will frequently be less inventory compared with period 10 in Dice Game 2.

There are two elements to variation – process variation and supply/order (or arrival rate) variation. This game illustrates the second of these. It is frequently an ignored element. The game is one justification for activities such as

- Stabilising orders
- Heijunka or production smoothing
- Regularising deliveries
- Mixed model scheduling
- The runner, repeater, stranger analysis when aimed at schedule stability.

Note: The advantage of reducing demand variation is more powerfully demonstrated in Dice Game 1

**Equipment**

4 pieces (products) per player, plus 30 pieces minimum in the upstream raw material store. One dice per player. Flip Chart.

# Dice Game 5: Kanban: The Good and Not So Good

**Key Areas**

A game that shows that Kanban is effective in limiting the build up of inventory, but can reduce throughput if there is large variation.

**Manufacturing or Service?**

Manufacturing

**Overview**

The game should be played in a session that includes Dice Game 2. In Dice Game 2, there is excessive build up of inventory. Kanban is a way to limit this buildup. In-process kanban squares limit the amount of inventory. Unfortunately, with high variation, they may also starve some workcentres, leading to a drop in throughput.

The game is played with Kanban squares between each workstation, and illustrates a simple Kanban pull system.

**Summary of Learning Points**

- Kanban is effective at limiting inventory build up.
- Kanban can limit throughput where there is high process variation – unless there is excessive inventory, but this more or less defeats the purpose.

**Players**

6 minimum; up to 11. With 12 or more, play two parallel games.

**Approximate Time**

30 minutes

**Description**

The game is played somewhat like Dice Game 2, with the players forming a sequential 'assembly line'. Pieces are passed along the line. A pull system is in operation. Players are only authorised to try to fill vacant spaces in the immediate downstream Kanban square, but may not overfill the Kanban squares.

**Instructions**

Players form a line as in Dice Game 2. A Kanban square with six spaces is placed between each player. Each player has one dice. At the start of the game all six spaces are filled in each Kanbansquare. This means that initial inventory is (number of players – 1) x 6. The player at the upstream end has an 'infinite' amount of inventory available – as in Dice Game 2.

Pull signals need to flow from the final downstream workstation up the line. This means that all players do not roll the dice simultaneously, but must wait for the next player downstream to complete his or her play first. (Note for steps below: 'he' could be 'she'!)

- The final (downstream) player rolls a dice, takes the lesser of his dice roll or the number of pieces available from the Kanban square immediately preceding his workstation, and hands them to the instructor (customer).
- The next upstream customer now rolls his dice. He is only authorised to fill vacant spaces on the Kanban square between him and the next downstream player. If he rolls a number more than the number of vacant spaces on the downstream square he may fill only the number of squares available. If he rolls less than or equal to the number of vacant squares, he may fill as many as his dice roll allows.
- He draws pieces from the next upstream Kanban square. If the number of pieces on the upstream square is more than or equal to the dice roll, he may use all that his dice roll (or the downstream Kanban square) permits. If the number of pieces on the upstream square is less than the dice roll, he may only take those pieces that are available from the upstream square.
- This sequence is repeated all along the line.
- The last (upstream) player rolls the dice and is authorised to fill but not to exceed the final Kanban square, if his dice roll permits.
- This completes the first sequence.
- 19 more sequences or waves are carried out. Go to the first step.
- Wait until the last upstream player has completed his throw before beginning the next wave or period. This is to avoid confusion.
- At no time should a kanban square contain more than six pieces. At the end of each period each kanban square should have at least one piece, or else an error has been made.

## Forms

Give each player except the final one an A4 sheet of paper. The players should then make Kanban squares, by drawing 6 circles on each A4 – big enough to place one piece (product) within each circle.

These Kanban squares are placed between each pair of players. At the start of the game the squares are filled with 6 pieces each. This is the upper limit of the number of pieces.

## Tables

As for Dice Game 2

## Expected results

After 20 periods, usually, in comparison with Dice Game 2

- The inventory is far less
- But, the output is also less.

## Instructor notes and discussion

- Explain the mechanics of the game.
- Ask players to predict the output. They will frequently not expect output to fall.
- Why does output fall? Because each player now has three variables, not just two as in the dice game. As well as the dice roll and the available inventory, players must also look ahead to the next square. A high dice roll may not take effect, because of look ahead. And starvation results from a cap on inventory coming through. The effect is communicated along the line.
- Kanban works very well in limiting inventory growth.
- But, where there is excessive variation, Kanban can limit throughput.

## Equipment

6 pieces of 'product' per player, plus about 30 pieces minimum in the upstream raw material store.
One dice per player.
One A4 sheet per player.
Flip Chart.

# Dice Game 6: Flexible Working

## Key Areas

- The good news and the bad news about flexible working
- A way to meet the target output with flexible labour
- Keeping track of input and output (?)

## Manufacturing or Service?

Any Manufacturing or Service operation having sequential workstations, flexible labour.

## Overview

The game starts with a challenge: achieve an output of 70 which was not possible in Dice Game 2. Players decide amongst themselves how to allocate capacity. It takes a lot of time, but most teams do achieve the target. The advantage of increased throughput must be set against the costs of flexible training, decision time taken, and perhaps most importantly, the fact that a wave-like system of throughput is established rather than good, steady uniform flow. Again, there are better ways. See later...

## Summary of Learning Points

- Increased throughput is balanced against cost of training for flexibility.
- Although the target can be achieved, much reorganisation and rescheduling is required.
- Decision taking involves time and cost.
- An undesirable throughput wave is established.

## Players

Same as Dice Game 2. Often done as a follow on.

## Approximate Time

1 hour

## Description

The game starts with a challenge: achieve an output of 70, which was not possible in Dice Game 2. But what is allowed is that the same number of dice as the number of players, can be distributed amongst any workstation or workstations, and can be varied from period to period. This represents flexible working.

For example, if you have 8 players, in period 1 there may be
- 4 dice used at workstation 1
- 4 dice used at workstation 2
- 0 dice used everywhere else

In period 2 there may be
- 2 dice used at workstation 1
- 4 dice used at workstation 2
- 2 dice used at workstation 3
- 0 dice used everywhere else

As in Dice Game 1, only the inventory available in the WIP squares may be used in any period.

The team of players must decide on the allocation. An allocation decision must be made every round.

**Instructions**

As with Dice Game 2, there is one workstation per player, a 'receive' square and a WIP square at each workstation, except the first workstation where, as with Dice Game 2, there is infinite inventory available.

The game has exactly the same rules as Dice Game 2 with respect to initial inventory (4 per workstation) and game mechanics – passing pieces along the line. The total number of dice is equal to the number of players (or workstations).

The instructor should call out the periods and record the output, if any.

Each period, players must decide how dice pieces are to be allocated between workstations. There may be considerable delay between periods, as the team decides on the allocations.

Use the same table as for Game 2.

**Typical Results**

Very often, the target of 70 is reached. A good way to play is to allocate most of the dice to the gateway (upstream) workstations, until a total of 70 pieces – less the starting inventory pieces of 4 per workstation – has been fed in. Then, push the pieces downstream in a wave, by allocating most dice to successive workstations each period starting upstream. Of course, do not use any more dice at upstream workstations where all necessary inventory pieces have been cleared.

**Instructor notes and discussion**

Some teams take time to realise how to play the game. Some teams do not reach the target because:

- they do not work out how to play the game
- they let in too little or too much inventory at the first workstation
- they miscount the pieces.

The instructor should comment on the amount of time taken, and on miscounting the pieces if this occurs.

Points to be made are:

- Flexible working can increase throughput or lead to achievement of target. This is due to reduction in the waste of capacity through having insufficient inventory.
- But the cost is in training and the time taken for decision making. Not free!
- More important, at period 20 the inventory has often been cleared out. This then is a problem for the next 20 periods. There will be no output for the next 6 or 8 periods. Customer service may suffer. To break out of this wave is a big problem!
- Amplification (Mura) results.
- Inventory record accuracy is important for best results.
- So, there are better ways….

**Equipment**

4 pieces of 'product' per player, plus about 30 pieces minimum in the upstream raw material store. One dice per player.
Receive and WIP squares for each player.

# Dice Game 7: Bottleneck and Uncontrolled Overcapacity!

**Key Areas**

- A warning on overcapacity and bottlenecks
- An optional introduction to the Drum Buffer Rope and CONWIP games.

**Manufacturing or Service?**

- Manufacturing, although many sequential service operations have bottleneck problems.

**Overview**

The game is very similar to Dice Game 2, but there is a clear bottleneck.

With large excess capacity upstream of the bottleneck, a large amount of inventory accumulates at the bottleneck.

Unexpectedly for some players, total output usually increases slightly above the average for Dice Game 2. Typical throughput is in the low 70's.

**Summary of Learning Points**

- Inventory accumulates at the bottleneck
- The accumulation of inventory at the bottleneck forms a buffer so that the bottleneck never loses capacity due to shortages.
- Downstream of the bottleneck, inventory clears out.

**Players**

As for Dice Game 2: one player per workstation; minimum of 6 workstations.

**Approximate Time**

20 minutes

**Description and Instructions**

All players are given two dice, except a player in the middle of the line who has one dice. This is, of course, the bottleneck. All players except the bottleneck add their two dice roll scores together. As with Dice Game 2, all players start with 4 pieces of inventory, except of course the first upstream workcentre.

The mechanics of the game are similar to Dice Game 2. 20 periods are played and the instructor records the total throughput and the total inventory.

Large amounts of inventory accumulate, so expect to have many pieces. Alternatively, later on in the game, occasionally remove 10 pieces of inventory from the bottleneck and ask the bottleneck player to remember how many pieces have been removed.

**Forms, Tables, Typical Results**

Forms as for Dice Game 2. Each player has a receiving square and a WIP square. Results are written on a flip chart as in game 2. There is no need to pause half way through the game.

Typical throughput is in the mid 70's, and ending inventory can easily reach 80 pieces.

**Instructor notes and discussion**

- After explaining the game mechanics, the instructor should ask the players to estimate the throughput after 20 periods.
- Write the estimates down and ask the reasons.
- Often, there is wide opinion from a very low estimate (because they have learned through earlier games about the dangers of variation), to very high (because of extra dice).
- Then ask the players where inventory will accumulate. (At the bottleneck.) What will be the effect of inventory at the bottleneck? (Capacity will never be lost.) So, what will the expected throughput through the bottleneck be? (About 70; depends on dice roll, not on shortages.) What will happen downstream of the bottleneck? (Inventory will often reduce because capacity is great – two dice.)
- Then play the game, and record the results.
- The good news: throughput is satisfactory.
- The bad news: inventory is excessive.
- Then ask for suggestions as to how it can be made all good. This is the lead into the Drum Buffer Rope game.

**Equipment**

As for Dice Game 2, but with two dice per player and around 80 pieces of inventory if no recycling is required.

# Dice Game 8: Drum Buffer Rope (DBR)

**Key Areas**

Effective flow control, and flow maximisation, when there is a clear bottleneck

**Manufacturing or Service?**

Mainly manufacture but some service operations have sequential operations with a bottleneck.

**Overview**

The game illustrates the effectiveness of the Drum Buffer Rope system – often used in Theory of Constraints scheduling. Like Dice Game 7, there is a sequence of players each with two dice, except the bottleneck which has one dice. The bottleneck is the 'Drum'. A 'Buffer' is set up in front of the bottleneck. An imaginary 'rope' connects the Drum to the gateway (first upstream) workcentre. Throughput is high, and inventory is low. This is a multi-stage Kanban system.

**Summary of Learning Points**

- The Drum Buffer Rope (DBR) system is a superior form of multi-stage pull.
- DBR works better than stage-by-stage Kanban where there is high variation.
- DBR can be applied at supply chain level, at plant level, and at cell level.

**Players**

As for Dice Game 2: one player per workstation; minimum of 6 workstations.

**Approximate Time**

40 minutes – with discussion

**Description**

The game follows on from Dice Game 7. Although not essential, it is recommended that Dice Game 7 be played immediately before this game. All players are given two dice, except a player in the middle of the line who has one dice. This is, of course, the bottleneck. All players except the bottleneck add their two dice roll scores together.

A Drum Buffer Rope system is discussed and set up before the game starts.

The mechanics of the game are similar to Dice Game 2. 20 periods are played and the instructor records the total throughput and the total inventory.

By comparison with Dice Game 7, throughput is maintained but inventory falls significantly.

## Instructions

Set up the game as for Dice Game 7. A workstation in the centre of the line is chosen as bottleneck. All players are given two dice, except the bottleneck that has one dice. All players start four pieces of inventory, except the bottleneck.

If two simultaneous games are run, locate a bottleneck at a different point in each game – perhaps positions 3 and 5 in a six player game.

The instructor explains the Drum Buffer Rope theory, and implements the system.

The mechanics of the game are similar to Dice Game 2. Dice are rolled, and pieces are passed along when the instructor indicates.

20 periods are played and the instructor records the total throughput and the total inventory.

## Forms, Tables, Typical Results

Forms are similar to Dice Game 2. The 20 period table should be set up on a flip chart. Typical results are a throughput of 70+ pieces, and dramatically less inventory than game 7.

## Instructor notes and discussion

- It is recommended that Dice Game 7 is played immediately before this game. If Game 7 is not played, certainly players must have played Dice Game 2.
- Explain the term 'bottleneck'. Strictly, a bottleneck limits the amount of throughput in a system.
- Tell the players that the 'Drum' is the name given to the bottleneck workstation, because it provides the drumbeat for the whole line.
- Explain that the 'Buffer' protects the bottleneck from ever wasting capacity because there is always sufficient inventory available. Ask what would be a good number of pieces in the buffer. Ask for the players' opinions. Generally less than 8 risks running out of stock. Why? Because the bottleneck may roll (say) two 6's in succession and the workstation before the bottleneck may roll two one's. Not very likely, but possible over a cumulative period. A good number is 10. If the group suggests 12 or more, accept this but suggest that the number above 10 should be set aside, and only used if necessary. This is good practice in the real world.
- Explain the 'Rope'. This is an imaginary link between the bottleneck and the gateway (first upstream) workcentre. The idea is that whatever the bottleneck lets out, the gateway lets in. So a communication link must be established. However, the gateway workcentre should still roll two dice each period and add the scores. In most periods the bottleneck (one dice) will roll less than the gateway (the sum of the two dice). If the bottleneck does roll more than the gateway, then the gateway player needs to pass on the sum of the two dice, but record the deficit. The deficit should be caught up over the next few periods, by passing on the bottleneck throw plus the deficit.
- In other words, the total inventory between the gateway and the bottleneck (the rope inventory) should remain the same, except if there is a deficit. Inventory at individual workstations may fluctuate. The instructor should check this after the first period, and half

way through the game. This inventory should be the bottleneck buffer + (the number of players between the gateway and the bottleneck but excluding these two players x 4). If this is not correct, a player has made a mistake. Investigate and fix before proceeding.

- Ask the bottleneck player to record any shortages when he or she is not able to process the full dice roll.
- Run the game and record the cumulative output, total inventory, and the rope inventory.
- Compare the results with Dice Game 7. Throughput will be similar but inventory will be much reduced. If comparing with Dice Game 2, throughput will be up, and inventory will usually be down. Dice Game 2 is not, of course, a direct comparison.
- Explain that Drum Buffer Rope (DBR) is a multi-stage pull or Kanban system. Pull should NOT be thought of as synonomous with Kanban. More correctly, a pull system places a cap on total inventory and responds to downstream demand. A push system has no cap, and work is launched by forecast or schedule.
- DBR is much more effective than stage-by-stage Kanban where there is high variation. Compare with Dice Game 5, if this has been played. Of course, Kanban can still be used to pull components into a line.
- In the real world, the buffer is a TIME BUFFER and needs to be regularly monitored. (The place to hold the morning production meeting?) In this game there is only one product so the buffer can be expressed in terms of inventory. But in real systems, the buffer needs to be expressed in time units. So maybe there are 20 product A on Monday in the buffer, but on Wednesday there are 15 product B. But, both represent (say) 6 hours of buffer.
- Explain that DBR can be used on three levels: At cell level: just like the game! Most cells have a bottleneck or DRUM stage. Why not, therefore, run a cell like a DBR system rather than attempting to balance the line? We saw the drawback of a 'balanced line' in Game 2. At plant level: the Drum may be a stage, a cell, or a workcentre. At the supply chain level: the Drum may be a complete plant. Either way, it works very effectively.
- Ask, what are the problems of DBR? Possible problems are the difficulty (sometimes) of locating the bottleneck and of shifting bottlenecks. These are partly addressed by the CONWIP Dice Game, next.
- Another issue is: how to introduce customer pull into a DBR system? This point is partly covered in Dice Game 10.
- A bottleneck can or should be used to regulate flow against demand. That is, fix the capacity of the most constrained resource to be in line with actual demand. This prevents overproduction. This works fine with stable demand. However, in a situation of increasing or decreasing demand, these trends need to be anticipated because it takes time to flow through from the bottleneck to the final workstation.

**Equipment**

As for Dice Game 2, except 2 dice per player. The 'Drum' (or bottleneck) player has only one dice. The pieces (products) could be lego bricks, nuts (as in bolts), or wrapped sweets.

# Dice Game 9: CONWIP

**Key Areas**

Effective flow control, and flow maximisation, when there is no clear bottleneck

**Manufacturing or Service?**

Mainly manufacture but some service has sequential operations with a bottleneck.

**Overview**

The game illustrates the effectiveness of the CONWIP (Constant Work in Process) system. Like Dice Game 7, there is a sequence of players each with two dice, except a bottleneck that has one dice. The bottleneck has no starting buffer and, during the game, the location of the bottleneck is changed. Throughput remains high, and inventory is low. CONWIP is a multi-stage Kanban system.

**Summary of Learning Points**

- The CONWIP system is a superior form of multi-stage pull.
- CONWIP is a good scheduling control system for shifting or uncertain bottlenecks. It is self-adapting.
- Buffer inventory tends to accumulate just where you want it to accumulate.
- CONWIP works better than stage-by-stage Kanban where there is high variation.
- CONWIP can be applied at plant level and at cell level.

**Players**

As for Dice Game 2: one player per workstation; minimum of 6 workstations.

**Approximate Time**

30 minutes – with discussion

**Description**

The game follows on from Dice Game 7 or 8. Although not essential, it is recommended that Dice Game 8 be played immediately before this game. All players are given two dice, except a player in the middle of the line who has one dice. This is, of course, the bottleneck. All players except the bottleneck add their two dice roll scores together.

The CONWIP system is discussed and set up before the game starts.

The mechanics of the game are similar to Dice Game 2. 20 periods are played and the instructor records the total throughput and the total inventory.

## Instructions

Set up the game as for Dice Game 7. A workstation near the centre of the line is chosen as bottleneck. All players are given two dice, except the bottleneck that has one dice. All players start four pieces of inventory, except the bottleneck.

The instructor explains the CONWIP theory, and implements the system.

The mechanics of the game are similar to Dice Game 2. Dice are rolled, and pieces are passed along when the instructor indicates.

20 periods are played and the instructor records the total throughput and the total inventory.

## Forms, Tables, Typical Results

Forms are similar to Dice Game 2. The 20 period table should be set up on a flip chart. Typical results are a throughput of 70+ pieces, and dramatically less inventory than game 7.

## Instructor notes and discussion

- It is recommended that Dice Games 7 and 8 immediately precede this game. If these have not been played, certainly players must have played Dice Game 2.
- Explain the term 'bottleneck'. Strictly, a bottleneck limits the amount of throughput in a system.
- A problem is 'shifting bottlenecks' or hard-to-identify bottlenecks. This may arise because of changes in product mix and/or machine breakdown. Note from Game 2: Accumulation of inventory does not necessarily indicate a bottleneck.
- In this game, locate the bottleneck (single dice player) towards the end of the line, but not at the end. At period 10, change the location to about one-third along the line.
- CONWIP (Constant Work in Progress) simply establishes a link from the end of the line to the beginning of the line. Whatever is let out, is let in. Hence work in progress in the entire system remains constant, even though inventory at individual workstations may vary.
- The CONWIP game does not have a buffer in front of the bottleneck, simply because the bottleneck is not known. Of course, in the game, we do know where the bottleneck is. To make the game comparable with the DBR game, give each player 5 pieces of inventory.
- Ask the players where they think inventory will accumulate. They will say 'in front of the bottleneck'. And where do you want it to accumulate? In front of the bottleneck! Now we will see that working...
- In the DBR there is a 'Rope'. Similarly in the CONWIP game there is a rope. (It is not called the rope, but it works in the same way.)  In CONWIP there is an imaginary link between the final (downstream) workcentre and the gateway (first upstream) workcentre. The idea is that whatever the final workcentre lets out, the gateway lets in. So a communication link must be established. The gateway workcentre should still roll two dice each period and add the scores. If the final (downstream) workcentre rolls more than the gateway, then the gateway player needs to pass on the sum of the two dice, but record the deficit. The deficit may be caught up by the gateway by passing on the final (downstream) throw plus any deficit.

- In other words, the total inventory between the gateway and the final downstream workcentre should remain the same, except if there is a deficit. Inventory at individual workstations may fluctuate. The instructor should check this after the first period, and half way through the game. This inventory should be the (the number of players – 1) x 5. If this is not correct, a player has made a mistake. Investigate and fix before proceeding.
- Ask the bottleneck player to record any shortages when he or she is not able to process the full dice roll.
- Output should be very comparable with the DBR game. Around 70 pieces. Perhaps a little lower due to the time that it takes to accumulate the buffer. But, the inventory buffer tends to accumulate just where you want it to – in front of the bottleneck.
- Point out that if the bottleneck has shifted upstream there will be a decrease in inventory downstream of the bottleneck. This may inflate the output somewhat.
- So, the CONWIP system is simple, powerful, and flexible. A problem is how to fix the CONWIP quantity. One way is to do this by trail and error. Start loose, with generous inventory, then decrease it slowly until a 'rock' is encountered and output falls.
- CONWIP can or should be used to regulate flow against demand. That is, adjust the flow into the beginning (upstream end) of the process to be in line with anticipated demand. This prevents overproduction or underproduction. It takes time to flow through from the beginning to the final workstation. So, CONWIP is an automatic adjuster against varying demand.

## Equipment

Two dice per player and 5 pieces of inventory per player plus (say) 30 pieces in the store at the upstream end. The pieces (products) could be lego bricks, nuts (as in bolts), or wrapped sweets.

## References and Further Reading

Hopp and Spearman, *Factory Physics*, Third edition, McGraw Hill, 2008 is the prime reference on CONWIP. Highly recommended.
For a more user-friendly version see Wallace Hopp, *Supply Chain Science.*

# Dice Game 10: OEE Implications and Limitations

## Key Areas

A game that explores effects of process availability, and warns against simplistic OEE

## Manufacturing or Service?

Manufacturing

## Overview

First, if necessary, the effect on throughput of improving availability as against lower availability but lower variation, may be demonstrated.
OEE (Overall Equipment Effectiveness) is widely used. One of the components of OEE is availability. Two machines may have the same average availability, but different breakdown and repair characteristics. The implications for inventory are large, especially when there is a bottleneck

## Summary of Learning Points

- Availability is important, but so too is variation. Conventional OEE does not look at variation, but it should do!
- Improving overall availability is key to improving throughput. But…
- The game shows why Mean Time Between Failure (MTBF) and Mean Time to Repair (MTTR) may be more useful than OEE for inventory and scheduling decisions.
- More frequent breakdowns but with rapid repair, generally require less inventory and have a shorter lead time than less frequent breakdown but with much longer repair time – for the same availability and throughput.
- Discussion on productivity vs OEE improves the understanding of OEE.

## Players

6 players minimum, each with two dice, in a sequential line, except the bottleneck that has one dice.

## Approximate Time

40 minutes

## Description

Optional: If players need convincing of the effect of availability and variation on throughput and inventory, run the standard Dice Game 2.

Explain that with average capacity (3.5) and no variation, throughput would be exactly 70 (20 x 3.5). Inventory would be unchanged.  In Dice Game 2, variation results in a decrease in output.

Each of the following games should be 20 periods minimum. When making a comparison with other games, like the DBR game, stop and review after 20 periods.

The DBR game should be run before running this game. In the DBR game, the bottleneck with a single dice has an availability of 3.5 / 6 = 58%

Then, the game is run as for Dice Game 8 (The DBR Game) with two dice per player and four product pieces per player. A bottleneck with one dice is located near the centre of the line. Place 10 pieces of inventory in the buffer in front of the bottleneck.

Overall workstation availability is maintained at an average of 3.5/6 = 58% but the type of availability is changed. The implications for throughput and inventory are explored.

**Instructions**

- Set up the game as for the Drum Buffer Rope (DBR) game. That is, two dice per player, but no dice at the bottleneck. Four pieces of inventory per player, except the bottleneck that starts with 8 pieces. 20 periods per game.

- In the first round of 20 periods, the player <u>at the bottleneck</u> does not throw a dice during the game. The player simulates various availabilities by using a pre-selected set of values or 'rolls' given in the next bullet point. All other players roll dice as in the DBR game.

- This game is run with the <u>bottleneck</u> having a repeating sequence of 6, 6, 6, 3, 0, 0. This has the same average availability of the average dice roll of 3.5 per period. All other players play as before with two dice.

- In the second round of 20 periods, the player <u>before the bottleneck</u> does not throw a dice during the game, but the bottleneck player does roll a single dice for each period as in the normal DBR game. The pre-bottleneck player simulates various availabilities by using a pre-selected set of values or 'rolls'in the next bullet point.

- This second game is run with the <u>pre-bottleneck</u> using a repeating sequence of (6, 6) (6,6) (4, 0) (0, 0). This has the same average availability of the average dice roll of 7 (2 x 3.5) per period. All other players play as before, with the bottleneck using one dice and all others two dice.

**Forms, Tables**

Players should use the standard inventory squares as in Dice Game 2.

The instructor should record the throughput results using the standard table displayed on a flip chart.

**Instructor notes and discussion**

For the first, or bottleneck game above, often inventory shortages develop at the bottleneck, most usually in the second or third sequence. If shortages develop, throughput will fall. Capacity lost cannot be regained. More buffer is needed. The gateway workstation will, however, let in a sequence of 6's, restoring the inventory but not necessarily at the right time. Even if shortages do not occur, waves of flow will develop further downstream. Make the point that, if customer orders are not in sync with these waves (and usually they are not), the consequences will be delivery shortfalls. In any of these cases, the situation is undesirable.

For the second, or pre-bottleneck game, capacity will not be fully used in the first sequence and in later sequences: 12 is thrown, but there are only 4 pieces of inventory in the first period. There is a fair chance that an inventory shortage will develop at the bottleneck, and throughput will fall.

Both of these scenarios illustrate how important the variation in availability is, not just the average as is measured by the availability element in OEE.

The instructor might wish to extend the discussion by asking the group to think of two situations:

In the first, availability and speed are both 100% but quality is 80%. In the second, quality and speed are both 80% and availability is 80%.

- What is the OEE in both cases? It is 80%.
- What is output per 100 minutes if the speed or rate is 1 unit per minute? It is 80 in both cases.
- What is OEE in both cases? It is 80%.
- But, what is the input in both cases – assuming a CONWIP or pull system is in place? It is 100 units in case 1, but 80 in case 2.
- What is the productivity in both cases? Productivity is output/ input; so it is 80/100 = 80% in case 1 and 80/80 = 100% in case 2.

This highlights that OEE may not be a satisfactory measure by itself!

**Equipment**

As for DBR Dice Game

# Dice Game 11: DBR with Backlog and Constraint Loops

## Key Areas

- Communicating changing demands to a line with a bottleneck
- Effective flow control
- Priority control

## Manufacturing or Service?

Mainly manufacture but some service

## Overview

This advanced game expands on the effectiveness of the Drum Buffer Rope system from Dice Game 8. Here two priority signalling systems are incorporated, that can be used to regulate flow and meet demand.

The concept may be used in manufacturing for a make-to-stock environment with uncertain demand. In some service situations there may be bottleneck stage in a sequence of operations, and at the same time large variation in in the arrival rate of work. For example, sales orders or insurance claims may be much greater at month end.

Unlike other dice games, in this game there is customer pull not just push to the final workstation.

This is a multi-stage Kanban or 'Rope' or Loop pull system. One loop monitors the buffer in front of the bottleneck, the other monitors the backlog of work arriving at the system.

See the figure 'Drum Buffer Rope with Priority'

## Summary of Learning Points

- A priority signalling system is an effective tool in an uncertain demand environment.
- A priority signalling system is an effective tool to regulate the control of the bottleneck buffer.
- Both signalling systems can be incorporated to form an adaptive flow control system.

## Players

As for Dice Game 8: one player per workstation; minimum of 6 workstations. The game may go on for more than 20 periods. The aim is to meet customer demands with minimal (or effective) WIP inventory.

## Approximate Time

40 minutes – with discussion.

**Description**

The game is set up as for Dice Game 8, with a sequence of workstations with a bottleneck workstation near the middle of the sequence. All workstations have two dice, except the bottleneck which has one dice. Periods, as usual, are under the control of the instructor. 20 plus periods per game.

This game is more of a demonstration than a game.
It is recommended to play the DBR game (Dice Game 8) before playing this game.

The game simulates a make-to-stock environment with uncertain demand. There is a finished goods buffer. Demands are communicated to the final downstream player. Two pull loops are established, one between the final workstation and the other between the bottleneck and the gateway workstation.

There are two buffers. One is located in front of the first workstation or gateway, the other in front of the bottleneck. Each buffer has zones of black, green, yellow, and red. The zone colours signal appropriate action (or no action) upstream.

**Instructions**

1. Set up the game as for the DBR game. Each player has two dice, except the bottleneck which has one dice.
2. The game is similar to the DBR game. There is a pull 'rope' between the bottleneck and the gateway, and a signalling rope from the pre-gateway buffer to the bottleneck. The gateway is the start of the process from where orders are launched into the system
3. An additional player should record the output from the last player, as in Dice Game 2.
4. Two buffers are established, each with zones of green, yellow, and red. These are given on the player cards. Note the maximum numbers of pieces allowed in each zone.
5. The buffers are FIFO lanes. At the end of each period the pieces are moved from left to right, remembering to not exceed the maximum number of pieces in any zone.
6. The gateway buffer begins with
   o four pieces in the green zone (full)
   o two pieces in the yellow zone.
   o nil pieces in the red zone.
7. The bottleneck buffer begins with
   o four pieces in the red zone (full)
   o four pieces in the yellow zone.
   o Two pieces in the green zone
8. The gateway player takes pieces from green to yellow to red. The zones are replenished by FIFO : inventory in yellow moves into green until green is full, red moves into yellow until full, new orders come in at the red end.
9. The bottleneck player takes pieces from red to yellow to green. The zones are replenished by FIFO: inventory in yellow moves into red until red is full, green moves into yellow until full, production flows in at the green end.
10. The bottleneck player needs to communicate with the gateway player as in the DBR game.
11. The instructor calls out the periods as in the DBR game, and the players, except the gateway and bottleneck players, play as in the DBR game rolling two dice every period.

12. The instructor takes the role of the customer and rolls one dice to generate demand requests. However, every 6 periods or so the customer should ask for 8 requests.
13. The instructor/customer places orders (pieces) in the red zone of the buffer/FIFO lane in front of the first workstation.
14. The bottleneck player rolls one dice and pulls from the buffer/ FIFO lane – first from the red zone, then if necessary from the yellow, then if necessary from the green zone.
15. The player immediately upstream of the bottleneck places fills the zones red until full, then yellow until full, then green.
16. Before the start of each new period, both FIFO lanes are replenished:
    o The pre gateway FIFO line is filled from green to yellow to red, but not exceeding the maximum number of pieces in each zone.
    o The pre bottleneck FIFO lane is filled from red to yellow to green, but not exceeding the maximum number of pieces in each zone.
17. Before the start of each period, check the status of each FIFO lane/buffer.
    o For the bottleneck, if there are fewer than 4 pieces in the red zone, there is a danger of insufficient buffer inventory. Perhaps there has been a problem in an earlier workstation. Investigate, and if necessary take remedial action by working an extra shift next time.
    o For the pre-gateway FIFO lane, if pieces (or orders) have accumulated into the red zone, a problem may have arisen. Perhaps orders have increased. Bottleneck capacity may need to be increased by working an additional shift.
18. Repeat for the next period, as from step 8.

**Forms, Graphs, Tables**

Display the usual table of 20+ periods on a filpchart. The customer (instructor) should also write up the demands and the deliveries made.

**Instructor notes and discussion**

- Players should have played the DBR game first, and be familiar with the game mechanics. If necessary, refresh their memory on the workings of the 'rope'.
- Take the first few periods slowly. Make sure that players understand.
- Possibly, before the game proper begins, have a trial period when a demand for 10 pieces is given to the gateway player. This should trigger appropriate action (i.e. an extra dice roll) from the bottleneck player.
- Explain that in many real world situations, it is necessary to monitor the buffers. The priority mechanism shown in this game is an effective way to keep on top of changing demands.
- The priority buffer system can also help adjust for breakdowns. This can be simulated by asking one of the players, usually other than the bottleneck, not to produce for one period.
- Remind players that in the real world, buffers may be inventory buffers or time buffers.
- Of course, if demand increases can be predicted, an extra shift can be laid on in good time.

**Equipment**
5 pieces (products) per player, plus 30 pieces minimum in the upstream raw material store. Two dice per player. Flip Chart. The pieces (products) could be lego bricks, nuts (as in bolts), or wrapped sweets. See the 'Drum Buffer Rope with Priority' Sheets.

# General Note on One Piece Flow Dice Games

The previous dice games (Games 2 to 11) are built around 'batch and queue' operations in as far as variable batches are made and moved forward to the next station. The following games enable you to play according to one piece flow, whereby as soon as a single part (or unit) completes a stage it is immediately moved forward and becomes available at the next stage, or of course goes into the queue in front of the next stage.

For many situations this is more realistic than the original batch games. In the classic dice games given earlier, there is an equal probability of any workstation capacity of between 1 and 6 units per period. In the Single Piece Flow Dice Games, instead of capacity per period the time taken for a piece at a workstation can be varied as required, and need not be uniformly distributed.

The one piece flow format allows players to explore some powerful concepts from 'Factory Physics' more easily than the batch and queue dice games. These concepts include the 'best case performance', feasible and infeasible regions, 'Critical WIP' level, and best WIP level for a line with variation.

The format of the Single Piece Flow Dice Game sheet enables an instructor to simulate a large number of cases, to suit the situation. For example, by rearranging the order of sheets an instructor can set up a game or simulation incorporating changeover and different workstation variability in any sequence. A TOC 'V' or 'A" plant can be simulated.

All the earlier games (Games 2 to 11) can be repeated using the game description given in Game 12 below. The lessons would remain the same.

# Dice Game 12: Basic Single Piece Flow

## Key Areas

- A game that allows simulation of a one piece flow line or cell with variation in station cycle times

## Manufacturing or Service

Mainly manufacturing. Some service relevance.

## Overview

Read the General Note on One Piece Flow Games if you have not already done so.

A group of players form an assembly line. For the basic round, all workstations have the same average cycle time of 1 period. Players roll a dice to determine the cycle time. As soon as a part is completed at a workstation it becomes available for processing at the next workstation.

## Players

5 minimum, up to about 12 maximum, each with a dice.

## Approximate Time

45 minutes

## Description

The game is played over 20 periods, but each period is divided into two, so there are actually 40 cycles. Even better is to play the game over 60 cycles or 30 periods.

The individual workstation cycle times vary randomly between 1, 2, and 3 half periods. The average is 2 half cycles or one full cycle. All workstations are similar.

After understanding that the average workstation cycle time is 1 period (2 half periods), players are given one unit (part) of inventory at each workstation. Players should be asked to predict the output from the end of the line after 20 periods.

Output is invariably less than the nominal 20 periods x 1 piece per period.

## Instructions

- Distribute one sheet of Single Piece Flow Dice Game (1) to each player.
- Give each player one dice. At every workstation, place one piece of inventory in the rectangle 'Inventory Queue'. Inventory can be Lego bricks (4 stud recommended), or washers or bolts (as in nuts and bolts).

- The instructor should prepare a flip chart with 40 half periods – numbers 1 to 40. There should be 3 columns on the chart: the half period number (1 to 40), the output from the end of the line during that period, and the cumulative output.
- Explain that the game will be played over 40 half-periods, equivalent to 20 full periods.
- Explain that the instructor will call out each half period. Players must wait for the instructor to announce the next half period before carrying out the actions for the half period.
- Explain that each workstation has possible but equally likely cycle times of 1 half period, 2 half periods, or 3 half periods. The average, of course, is 2 half periods or one full period.
- Each game card has three rows, with one, two and three squares. These are not three parallel machines but simply a way to record progress on one process having three equally likely times.
- A process should never have more than one unit (total) of inventory in any process square. Emphasise this!
- Ask the players to predict the output from the end of the line. Of course, 20 would be the 'rational' guess. Alternatively the instructor could say that the target output is 20 units.
- Tell the first player to always have one part (only) in the Inventory Queue box at the start of each period. Draw parts as needed from the box of Lego parts as necessary.
- The game begins: the instruction steps are:
  1  The Instructor will call the next period.
  2  If there is a part in a process square, move the part one square to the right.
  3  If the part is completed, move it to the Inventory Queue area of the next player.
  4  If there is still a part in the process, stop and wait for the next period.
  5  If there are NO parts in the inventory Queue, stop and wait for the next period.
  6  If there are NO parts in the process and there is at least one part in the Inventory Queue, roll the dice.
  7  If you roll a 1 or 2, move the part into the first Time = 1 process square; if you roll a 3 or 4, move the part into the first Time = 2 square, if you roll a 5 or 6, move the part into the first Time = 3 square.
  8  Note: The first operator should always keep one part in the Inventory Queue and then follow steps 3 to 8.
- Initial setup of game is one part in the Inventory Queue square of each player.
- Repeat for 40 half periods.
- The player at the end of the line should inform the instructor if any part is complete.
- The instructor will record the output for each half period as either 1 or 0, and complete the cumulative output column.

### Equipment

- One dice, one 'One Piece Flow Dice Game (1) for each player.
- Lego or Nuts (as in nuts and bolts) to be used as parts. A minimum of 2 parts per player

### Notes to Instructor

These notes are very similar to the Dice Game 2 Instructor Notes.

Tell the players that nominal throughput (output) of the line is 20 x 1 = 20 parts. Ask the players to estimate the total output from the line. Write these estimates down on the flip chart.

<u>Half way through the round</u>, after 20 half periods, stop and review. Point out that the output is less than the nominal 10. (On rare occasions it may equal 10.)

Also record the inventory in the WIP squares of each player. Ask the players to notice that the equally distributed 1 part for each player that the game started with has now become varied – some with 0, some with more.

Add up the total inventory in the game. (The sum of all parts in the WIP and the process squares.) Very likely, this total will exceed the initial quantity of 1 x number of players. Point out that inventory has increased.

Now do the 'Deming' charade. Go along the WIP squares. Find out who has the most inventory. Say 'That is not good enough. You will have to improve! I will be watching you!' Then find the person with least inventory in the WIP square. Say, 'Great performance. You are just the sort of employee we are looking for! Well done! Put it there!' (Shake his hand.)

Then, do the second 20 half periods as for the first 20 periods.

After 40 half periods:

Point out that

- The target of 20 has not been reached. It is never reached! Why? Because of variation, not only at each workstation, but also because of variation higher upstream. Goldratt calls this 'statistical fluctuation and dependent events'. A general conclusion is that OUPUT < NOMINAL CAPACITY
- Inventory has increased. It always does! Why? Because what is being let in is an average of 1 per period but the rest of the line together is restricting flow to below 1.
- Variation is the enemy! If there was no variation – say everyone had a dice with only a 3 output would have increased and inventory would have stayed the same.
- So, with variation, if you expect to reach your target, you will frequently be disappointed. Many managers don't realise this fundamental point!
- Accumulation of inventory does not necessarily indicate a bottleneck. It may be too soon to judge. Over the long term however, accumulation may well indicate a bottleneck.
- Deming: often, as in the example, the player who was given a warning has improved. Clearly, then, management warnings and motivations work, don't they? So management should get a bonus! Clearly, this is rubbish. But how often does that happen in a company. Deming was fond of saying that many managers don't understand variation and are apt to hand out undeserved rewards and punishments. Of course, this is the difference between common cause and special cause variation. Here we have common cause or natural variation. Rewards and punishment is not appropriate. (And, perhaps the player who was 'the best' is no longer so. Sometimes, as Deming said, praise has 'gone to his head'.......)
- The problem of 'balancing a line' (Goldratt says 'Balance flow, not capacity'. Here we have a perfectly balanced line, but it fails to produce the target output. Much effort is put into balancing real lines or cells. But, as illustrated here, this may be futile if there is much variation. A balanced line can only be effective if there is minimal variation. It is not that balanced lines don't work, just that they don't work well with variation. There are better ways....
- The longer you play the game, the more inventory there will be. Eventually, after a long period, there would be so much inventory in the system, and so much inventory in front of

each work station, that the effects of variation will be buffered by inventory. The game will then reach equilibrium or 'steady state' with average input being matched by average output. In this case, then, we would have Little's Law. That is WIP = Throughput x Lead time.

- Once again, a *Law of Factory Physics* from Hopp and Spearman is *"Variability in a production system will be buffered by some combination of:*

    o *Inventory*
    o *Capacity*
    o *Time"*

So, if you have variability you will have to pay for it by extra inventory, capacity, or time. There is no way out of this! But it is a strategic choice as to whether you elect to have more inventory (to buffer demand and process variation), or more capacity but less inventory, or to make your customers wait thereby enabling operations to be leveled and capacity and inventory to be reduced.

Another true statement is "As soon as you have variation, someone or something will wait" : Inventory, Customers, Workers, or Machines.

# Dice Game 13:  Single Piece Flow: Drum Buffer Rope and CONWIP

These games are similar to Batch Dice Games 8 and 9  except, of course, that Single Piece Flow Dice Game Sheets are used.  Use Single Piece Flow Dice Game (12) at all workstations except at the chosen bottleneck workstation.

Single Piece Flow Dice Game sheet (3) can be used to create a mild bottleneck at a workstation somewhere near the middle of the line.

Single Piece Flow Dice Game sheet (4) can be used to create a more severe bottleneck at a workstation somewhere near the middle of the line.

Run the games using the guidelines and instructor notes for Dice Games 8 and 9.

For the Drum Buffer Rope and CONWIP rounds, there is a pull signal (or authorisation of the first player to start work) only when a part comes out of the bottleneck (DBR round) or final process (CONWIP round). In other words, the first player should begin making another part only when a part is completed by the bottleneck player or final player respectively.

If you have extra time, you can incorporate BOTH Single Piece Flow Dice Game sheet (3) AND Single Piece Flow Dice Game sheet (4) in successive workstations. Other workstations should use the sheet for Single Piece Flow Dice Game sheet (1). Now you have a bottleneck where Single Piece Flow Dice Game sheet (4) is used but a near bottleneck where Single Piece Flow Dice Game sheet (3) is used.  This can lead to some interesting results – a shifting bottleneck and a starved bottleneck may occur.

# Dice Game 14: Critical WIP and Best Case Flow

**Key Words**

Critical WIP, Best Case Performance, Bottlenecks, Variation, Factory Physics

**Manufacturing or Service?**

Manufacturing

**Description and Discussion**

Consider an assembly line with five stations. The cycle time is 1 period at each station. No variation. Now run a CONWIP system, where as one unit of production is completed another is started – so input and output are matched exactly. Units are passed onto the next stage without delay.

For this initial round, ignore variation. No dice are rolled. All jobs pass through the centre line squares of each station.

Remember, throughout all these single piece games, that there should be no more than ONE part in any station square.

Start with one unit of inventory. It would take 5 periods (10 half periods) to complete all five stations, so throughput would be 1 unit every 5 minutes. But only one station would be busy at a time. Lead time is 5 periods.

Now repeat with two units. Throughput would be 2 units every 5 periods. Two stations would be busy at any time. Lead time is 5 periods for each unit.

Repeat again for each of 3, 4, and 5 units of inventory With 5 units, all stations are occupied. Lead time remains at 5 periods. Throughput is 5 units in 5 periods, or one per period. Use CONWIP: in other words as soon as a unit is completed at the last station another unit can be started at the first station.

But when 6 units are in the system, something interesting happens. Throughput in 5 periods does not increase, but stays at 5 units. But leadtime increases to 6 periods because there is one unit not being worked on.

And for 7 units, throughput remains at 1 per period, but leadtime increases to 7 periods. And so on. See the Game Sheet.

So at 5 units of inventory, throughput reaches a plateau but leadtime is at a minimum. This is the CRITICAL WIP. Notice that critical WIP = bottleneck rate x sum of process times = 1 unit per period x 5 periods = 5 units. (Or, using the half period squares on the sheet, ½ x 10 half periods = 5 units.)

So the 'best case performance' of any line occurs at the critical WIP level when there is no variation.

The general case has the same formula for critical WIP. If there are, for example 5 stations with cycle times 3, 4, 2, 3, 3 minutes, then the bottleneck is the 4 period station, with bottleneck rate 1/4. Sum of process times is 18. Critical WIP = 15/ 4 =  3.75 or 4 units

A simulation to illustrate the general case of critical WIP, and the formula for critical WIP = BNR * sum of process times, can be tried by using the Single Piece Flow Dice Game sheets (1), (3), and (4) but ignoring variation. In other words, all dice rolls are considered to be a 3 or a 4, so that the process always goes through the centre line on the process sheet. The bottleneck rate will be 1/4 per period (or two half periods), and the sum of the process times will be (the sum of the central line boxes for all process sheets used in the game.  Try it and see; it is interesting, easier to see than explain!

Now repeat the game with variation. That is, get players to roll the dice and select one of three process times.

If you have time, run 4 games with 3, 4, 7, 9 units of inventory respectively, each with 6 players using CONWIP. Remember, CONWIP lets in a unit of inventory at the start only when a unit of inventory is completed. In other words, the first player should begin making another part only when a part is completed by the final player. Hence CONWIP or constant work in progress.

 If you don't have enough time run one game with 5 units of inventory, 5 players, and CONWIP.

You should get results similar to that shown in the graph.

**Instructor Discussion**

- Once again, variation is the enemy.
- Critical WIP is an easy thing to calculate for any sequential assembly line. No line should be run with less than the critical WIP.
- Critical WIP is 'the WIP level that achieves the maximum throughput and minimum lead time in a process with no variability' (From Hopp, 'Supply Chain Science')
- With variation, as inventory increases throughput slowly increases. But very large amounts of inventory are required to achieve output near to the bottleneck rate.
- There are two ways to address this problem:
    - Protect the bottleneck with buffer inventory, and establish a Drum Buffer Rope pull system
    - Run the line with a CONWIP system, ensuring that there is somewhat more inventory than the critical WIP level, and less than (say) twice the critical WIP level. This latter version is shown on the graph.
    - The greater the variation, the greater the need for inventory to maintain throughput. The upper bound of the Lean zone will be higher.
- In any real assembly line it is relatively easy to establish the bottleneck rate, the critical WIP level. Then plot your actual position in terms of throughput and WIP. Do you match up? Are you well outside the Lean zone?
- Little's law can also be demonstrated. It is WIP = throughput rate x lead time.  For the no variation case, throughput is 1 per period, lead time is 2 periods, so WIP is two units.  For the variation case, throughput is ¼ and lead time is 15, so WIP is 3.75 or 4 units.

The Lean Games and Simulations Book

# Dice Game 15

An additional game sheet is included to allow inclusion in the simulation of a batch process such as a chemical process, or press requiring a setup.

In this case, inventory needs to accumulate before the batch process can begin. Of course, this introduces a delay and an inventory buffer is required after the process in order not to starve the line.

The instructor can alter the length (or size) of the batch by deleting any number of squares in the process.

This game sheet can be used with the other dice game sheets in a large variety of simulations. For instance, a TOC 'V' plant, and an 'A' plant can be simulated.

# Kanban Game

## Key Areas

Kanban, Pull, Pull Signals, Kanban loop, Supermarket

## Manufacturing or Service?

Widespread application in manufacturing. Some service.

## Overview

A series of short games to illustrate kanban and the number of kanbans required in situations of increasing complexity.

The use of the kanban card formula is developed.

## Summary of Learning Points

- 'Kanban' is a word for 'signal'.
- Kanban is a pull system responding to demand.
- Kanban cards circulate around a kanban loop.
- As cards go around a loop they are in one of 4 situations:
    - At the point of use (or 'supermarket') with a container of inventory
    - Travelling to the feeder operation or supplier as a pull signal (an empty container can also act as a kanban)
    - At the feeder station (which may also be a 'supermarket') waiting for the container to be filled or to be moved
    - En route back to point of use accompanying inventory
- There may be four types of inventory in a kanban loop:
    - Batch inventory to cover batch demand
    - Cycle inventory to cover demand during the signal and delivery phases
    - Buffer stock to cover customer demand variation
    - Safety stock to cover manufacturing variation (breakdown, defects)
- A 'supermarket' in a kanban loop is a temporary storage location where on operation 'goes shopping' to obtain inventory
- The pull signal is triggered by a container becoming empty, or when the first part is removed from a container. The former is preferable.
- Kanban is suitable for repeating demand.

## Approximate Time

75 minutes

## Description

- The group is divided into teams of 5 players each.
- The game is set up for teams to compete. Each round, the best team is the team with the smallest number of containers in circulation at the end of the round. If a team is unable to satisfy a customer demand in any period it is disqualified.
- A 10 second period increment display should be available.
- 4 stud Lego bricks or wrapped sweets are the products. About 50 per team is required.
- See the Kanban Game Sheet for player activities.
- A Round:
    1. The instructor walks through a number of period pull cycles as a demonstration. Start this round with 15 products (3 containers of 5) in the pre-Production supermarket, and (say) 20 products. Each period the customer asks for 5 products. The Production operation removes 5 products from a cup (container) and gives them to the customer. The Production operation places the empty container in the kanban collection point. The material handler will take the empty container to the feeding operation, dropping off the empty container at the beginning of the next round. The feeding operation fills the container. The filled container is delivered back to the Production supermarket at the beginning of the next period. The instructor should then explain the kanban formula. For the first round, demand is 1 container, lead time is 2 periods, probably good to have one risk (or safety) container in the supermarket, container quantity is 5 (or 1 container). Hence No = $((1 \times 2) + 1) / 1 = 3$ containers.
    2. Teams must now decide how many containers are required if there is one product per container, not 5.
    3. How many containers when variation is introduced by having the customer roll a dice? Demand is 4, 5, 6. (roll 1=4, 2=5, 3=6). Buffer stock is required.
    4. How many containers when there is a 'runner route' or tugger system where cards are collected at regular but spaced intervals?
    5. How many containers when batches of 25 are made at the feeder station, taking 30 seconds to make a batch?
    6. How many containers when there are two products in the loop, and where batches of 25 are made at the feeder station?
- Discussion and calculation is required between rounds

## Players

5 per team. More than 4 teams is not recommended.

## Player Instructions

Refer to the Kanban Game Sheet.

## Instructor notes and discussion

- Kanban is, of course, one of the most versatile and well-known tools of Lean. It is one way in which a pull system can be made operational.
- Despite this, there is widespread uncertainty about kanban calculations. This game gives practice.
- Some organisations (Toyota) use both production kanbans and withdrawal (or move) kanbans. This distinction, used for 'fine tuning', is not used here.
- The visibility of the system is an important factor. Point this out.
- Be prepared to demonstrate Round 1 slowly and patiently. To use the TWI phrase 'Continue until you know that they know', and 'If the worker hasn't learned, the instructor hasn't taught'.
- Be prepared to talk players through the kanban calculation.
- Each round must be played as an ongoing situation. In other words players must not stop production in anticipation of the end of the game.
- Containers cannot be removed during a round.
- When a full container is delivered to the Production supermarket, it is immediately available. In other words it can be used to supply the customer during that same period.
- Limit discussion between rounds to 10 minutes
- Have a flip chart showing results for each team, each round.
- Bricks or sweets can be recycled during a round.
- You can use an 'After Action Review' after each round.
- You can use Kata principles between rounds. The target condition is uninterrupted flow, with no demand failures and minimum inventory. What is then preventing the team from having just sufficient inventory available to meet demand?

## Equipment

For each team:
- 40, 4-stud bricks, nuts (as in nuts and bolts), or wrapped sweets
- 15 paper cups (containers)
- a period timer (better to use a data projector and PowerPoint display)
- a count-down timer for the batch cycle
- copies of the slides

## References and Further Reading

- John Bicheno and Matthias Holweg, *The Lean Toolbox*, 4th edition, PICSIE, 2009. Here we show kanban as one of several integrating concepts in a full manufacturing system. Types, advantages and disadvantages are discussed.
- Art Smalley, *Creating Level Pull*, LEI, 2004. This shows how kanban can be combined in general value streams.

# CONWIP and Drum Buffer Rope Advanced Explorer Game

**Key Areas: Pull systems, kanban, drum buffer rope (DBR), constant work in process (CONWIP)**

**Manufacturing or Service?**

Manufacturing. There has been limited use in service.

**Overview**

This series of exercises or games illustrate some of the alternatives that can be considered in using the DBR and CONWIP approaches.

The classic Drum Buffer Rope (DBR) and Constant Work In Process (CONWIP) approaches are multi-stage pull systems. They have many advantages over stage-by-stage kanban, especially in more complex environments.

In practical situations, the signal period for pull may not be instantaneous, the pull quantity may not be exactly equal to the exit quantity because of batch reasons, and there may be priority issues.

This game allows players to explore some scheduling possibilities. The game should not be played until prior scheduling dice games as described in an earlier section have been played or understood.

The game is more of a demonstration than a game, but the group of players can run their own alternatives after the initial two rounds.

**Summary of Learning Points**

- DBR and CONWIP are powerful approaches, but there are detailed alternatives in operating these systems. Both have the considerable advantage of limiting the total inventory.
- Signal frequency, batch sizing, and job priorities should be considered.
- Every variant in both DBR and CONWIP has advantages and disadvantages.

**Approximate Time**

   1 hour

**Description**

- The game is played over four rounds. Several parallel games are possible.
- Set up an 'assembly line' of sequential operations as in the dice games.
- The game should be played after the DBR and/or CONWIP dice games. The player rules in those games apply here. Please refer to Dice Game 9.
- Make four copies of the Order Cards for each team. Laminate if possible. Cut out the cards. Shuffle the order cards.

- The pile of shuffled cards represents the order in which demands will be received each period. There is no possibility of sorting the demands before they have been received.
- For each round, from the top of the pile of shuffled cards, distribute 4 cards to each player starting at the upstream end of the line. When the cards have been distributed, the remaining cards are given to Process 1 (Player 1).
- Each player has one dice, and rolls once per period, drawing on the inventory cards available in front of the process at the beginning of each period. Products should be processed by each player in a FIFO manner.
- For simplicity, in each round, Process 1 (player 1) does not roll a dice but merely launches the same quantity as the signal from the final process indicates. In other words, it is a CONWIP loop where, each period, the number of jobs completed by the final process is the number of jobs started by the first process.
- At the end of each period, the final (downstream) player should write the period number on all order cards that were completed by him or her during that period. Write boldly on the card but do not obscure the due date.
- After the 20 periods in each round have been completed, record for each team:
  - The number of products completed
  - The work in process inventory; this should be unchanged from the initial initial inventory.
  - For each completed product, calculate the periods early or late. This is the completed period written on the card minus the due date printed on the card. Postive indicates early, negative indicates late.
  - Add up the sum of the periods early or late for all completed jobs.
  - Draw a histogram of days early or late.
- At the end of the round, collect up the WIP cards in order beginning at the downstream end. Place these back on the top of the pile of cards. The final process should maintain the order in which products are completed. Place these on the top of the pile. (Therefore the sequence of order cards will be approximately the same for all rounds.)
- The four rounds are shown on the Game Sheet and are as follows:
  1. This is the basic benchmark round as in Dice Game 9. Use two products. Run the round over 20 periods. The pull signal is communicated to Process 1 from the last Process every period.
  2. In this round, we imagine that, for practical reasons, the pull signal can only be sent every two periods. The last process should therefore add the number of products that are completed every two periods, and communicate this quantity to Process 1. Process 1 launches products every 2 periods, taking the release quantity from the pile.
  3. In this round we imagine that for process reasons batches of similar products must be launched. This round is similar to round 2 except that the release quantity is determined and then products are sorted into batches of similar products before being launched. Product A, then Product B. Ignore due dates.
  4. In this round attention is given to due dates. This round is similar to round 2 except that the release quantity is determined and then products are sorted by due date before being launched; Product A, then Product B.

**Players**

Each team should have a minimum of 6 players, and maximum of 10.

## Player Instructions

- Use the basic instructions as for Dice Game 2, but set up a CONWIP loop as in Dice Game 8.
- The instructor should distribute the Game Sheet.
- The instructor should explain the four rounds above, round by round.

## Forms, Graphs, Tables, Typical Results

- Game Sheets
- Give a waterbased felt pen to the final (downstream) player if cards have been laminated.
- If cards are not laminated, give the final player a thick felt pen.

## Instructor Notes

- CONWIP and DBR are increasingly popular multi stage pull systems, having several advantages over stage by stage kanban.
- In real situations it is frequently not possible to run an exact DBR or CONWIP system. Practically, the work content between jobs may vary and the frequency of the pull signal has to be decided.
- Between pull signals jobs with different due dates accumulate. There is then an issue of whether jobs should be pre-sorted by due date before launching.
- If jobs are sorted by due date, over what period should the sorting take place? In other words by launching jobs in FIFO order may delay subsequent urgent jobs.
- Likewise there may be changeover (or setup) issues. Should like jobs be grouped, favouring reduced setup but possibly prejudicing due date performance?
- The group can invent their own variants – for example changing the signal frequency and pre-sorting jobs by either due date or product type or both.
- By the way, it is quite possible to have more than one DBR loop or a CONWIP loop in an end to end system.

## Instructor notes and discussion

- This game is labeled as DBR and CONWIP. Strictly it is a CONWIP game, but the mechanics and considerations are similar for both.
- The game extends the basic dice game that uses only one product without product due dates.
- Round 1 is a more realistic situation than the basic Dice Game 8 or 9.
- Round 2 is even more realistic. Here waves go through, particularly upstream. Holding more WIP inventory can help, but at a penalty of cost and lead time.
- Round 3 may be necessary for processing batch reasons. Of course the first batch performs better, the second batch worse. If product contributions or profit are different, launch the higher contribution batch first. Due dates for the second batch suffer.
- Round 4 allows due date sorting. Of course this improves due date performance. In practice, there is a question of how many periods demands should be aggregated before sorting. Shorter periods: maybe all about equally late? Longer periods: some early, some very late?

**Equipment**

- 4 order cards sheets per team. Preferably laminated. Cut out the order cards.
- One dice for each player, except Player 1, in each team.
- Flip chart for results.

**References and Further Reading**

- John Bicheno and Matthias Holweg, *The Lean Toolbox*, 4th edition, PICSIE, 2009.
- The classic easy-to-read book on DBR is Goldratt and Cox, *The Goal*, North River Press, 1986
- Wallace Hopp, *Supply Chain Science*, Waveland, 2008 (for CONWIP)
- David Anderson, *Kanban*, Blue Hole Press, 2010. This discusses the use of kanban and DBR, mainly in service situations.

# The Cut and Fill Games

The set of 'cut and fill' games is a more advanced alternative to the Airplane Games. In each game, players colour in and cut shapes – hence the name.

There are five games in the set:
- An initial Push system.
- The Line Balance Game gives practice of traditional line balancing, but in a mixed mode environment.
- The Line Balance Game extends into a Kanban game, and a CONWIP Game.
- Finally, 'Bucket Brigade' line balancing can be demonstrated in a slightly more complex environment than the Airplane Game.

**Keywords**: Line balance, cycle time, takt time, mixed model, cell design

**Times**: 30 to 40 minutes for each cut and fill game, except the Game 2 where data collection and analysis will extend the time to 1 or 1.5 hours.

## Manufacturing or Service?

Manufacturing. A small number of service organisations operating in a repetitive transactional environment would find the game relevant.

## Overview

This is a series of games for learning cell design mechanics. The games are more complicated and advanced than the airplane games in that there are several products with a variety of operation cycle times.

A line is set up. In the first round this may be spread out all around the room. Work proceeds with a push system for 10 minutes. The quantity completed is measured.

Operation cycle times need to be measured. The line is balanced against a percentage of takt time. Then re-run with Kanban, CONWIP, and Bucket Brigade line balancing.

## Players

10 minimum per game for all games.  It is possible to run two simultaneous games in parallel.

## Materials Required

The following are the requirements for each game. Double the quantities where two games are run in parallel.

- About 25 copies of the Cut and Fill Squares sheet; batches of 4 squares must be cut out from 20 sheets, and batches of 1 square cut from 5 sheets.
- 4 pairs of scissors
- 4 thick black felt pens
- 3 of each felt pens: red, blue, green

- Glue stick (e.g. Prit stick)
- 1 count down timer. A watch might do.
- One copy of Cut and Fill Player card for each player.
- One copy of Operator Balance Board for each player (Game 2 only)
- Three copies of Player Instructions (to be shared)
- One copy of Flow Diagram
- One Copy of Cut and Fill Balanced (?) sheet

In later games
- 8 copies of Kanban Square
- 2 copies of FIFO Lane

## General Instructions for all Cut and Fill Games

- A batch quantity of a product is made up by cutting out a batch quantity from a Product square strip (this is a number of blank squares), adding the appropriate shape or outline, colouring in as appropriate, writing in the product number, cutting out the shape, glueing the shape to a blank quare, and inspecting.
- See the Flow Diagram figure, and the Player Card. There are eight operations and six products. The eight operations are given on the Player Card. The six products (numbers 3 through 8) have respectively 3, 4, 5, 6, 7, 8 sides.
- In each game the target output is 40 completed products in 10 minutes.
- In the first round, you may choose to spread the various stations around the room. In this case a material handler will be required. Alternatively all rounds can be arranged in a line or U shaped cell. A U shape gives greater flexibility for balancing.
- Make sufficient copies of the Cut and Fill product squares. 30 pages for each game (total 4 x 30 = 120 for all games) should be more than adequate. Cut out batches of 4 squares as indicated and leave a pile of 15 in front of Station 1. Also, place a pile of 15 x 4 squares at Station 7, and ask player 7 to cut out individual blank squares before the game begins. Alternatively, the instructor can cut these out beforehand.
- See the Materials Required table. To be safe, double up on these quantities because in later rounds some operations are done in parallel.
- Give a Customer Order Sequence sheet to Player 1 and to Player 2. Player 1 needs the sheet to determine the batch size. Player 2 needs it to determine the product type to be made.
- The Player Instructions are simple. They are given on one sheet. It is not necessary to make an instruction sheet for each player. Four copies per game is adequate.
- Check that Player 1 understands that he or she must cut out the batch quantity as given on the schedule, and must not cut out individual squares. Where the batch size matches a left over (previously cut) quantity, this can be used.
- Check that Player 3 understands that he or she needs to time 10 seconds whenever a colour changes. During that time no work may be done on the product. This is typical in real life where, for example, there is a colour change or fixture change.
- Check that Player 4 understands that a colour-in is not required in the case of products 4, 5 and 7. These products can bypass operation 5.

**Priming the Line**

Prime the line with two of Product 3.
In order to make comparisons between the various games the line needs to be 'primed', in EVERY game. That is, each station needs to start with inventory to work on when the game starts. The instructor should give TWO products (not a batch or strip of four products) to each station 2 to 8. Use Product 3 – the triangle with vertical line. Cut out 7 x 2 product squares. Each station should get two part completed product squares appropriate to that station. In other words, Station 2 gets two blank squares, Station 3 gets two squares with the triangle and vertical line drawn, Station 4 gets two squares with triangle and coloured in red, and so on.

**Efficiency and Productivity**

In each game the productivity and efficiency should be measured.

> Productivity = output / input
> = number of completed products / number of players

> Line Efficiency = (sum of individual station times)/(players x 15)

Later the takt time will be calculated at 600 / 40 products = 15 seconds.
The line will actually be balanced at 12 seconds, but this of course represents a loss of efficiency. Only if perfectly balanced, will line efficiency be 100%

In Lean, however, we are less concerned with line efficiency but more concerned with meeting the schedule, and productivity. As Goldratt says 'balance the flow, not the work'. A perfectly balanced line is a near impossibility.

# Cut and Fill Game 1:  Unbalanced Push

Run the game as a push system. Start with no inventory in the system, one player per station. Each player simply pushes the batch (or order) along to the next player. Player 1 starts each new order as soon as he or she has finished the first operation.

Batches are pushed, not individual products. This is why station 1 should not cut out individual squares but instead cut out the required batch quantity.

Stop after 10 minutes. If you have an extra player ask him or her to time the lead time for Order number 4 (Product 5).

In this first game products stay together in their batch sizes until they are cut out at station 6.

Say that the target is 40 products in 10 minutes. Time the game and stop after 10 minutes.

Record the game performance on the Measures Sheet.

Discuss the game with the group. They will have noticed that there are surges of products going through the line, caused both by the batches and by the different product task times.  Stage 6 (cutting out the shape) is a particular bottleneck, especially when cutting out the arrow product.

# Cut and Fill Game 2: Line Balance with Push and One Piece Flow

First, it is necessary to time each operation : 1 to 8. Of course, there is the possibility of 8 operations x 6 products = 48 operations! Also, there is variation when making the same product. So there is variation between products and within products.

This is the reality of more complex assembly line operations: there are too many different operations to time individually. The group can discuss this point.

Here, a reasonable solution (not the only one) is to select one product of 'average' work content, say product 5, and time a batch of four at each of the 8 stations. Then record the cycle time of each operation by dividing by 4. Do this exercise. If you have a proper split timer, for example on an iPhone, it is instructive to record the time of each individual product on each individual operation (4 x 8 = 32 times). Then look at the variation.

Another factor, important in some assembly repetitive tasks, is 'learning' or 'the learning effect' or 'learning rate'. The more often you do a job, the better you get at it and the less time you take. The 'learning rate' is usually expressed as a decimal or percentage that tells by how much times will reduce for each doubling of units made. An 80% learning rate means that if it takes an average of 10 minutes for the 10th unit, it will take 0.8 x 10 = 8 minutes for the 20th unit, and 0.8 x 0.8 x 10 = 6.4 minutes for the 40th unit, and so on. In this game learning does play a role, but it will be ignored.

Fill in the average activity times on the Balance Table. By adding the times together, the line efficiency can be calculated.

Now calculate the takt time. 10 minutes x 60 seconds = 600 seconds / 40 products = 15 second takt time. But, recall the first dice game! Variation means that to balance against a takt of 15 seconds will be a disaster. So, balance agains what Toyota calls ' planned cycle time' which is a percentage of takt.

Here, we will take 80% of takt or 12 seconds.

A note on this: for new cells, be conservative. Start with a low percentage, like 80 to 85% and then move it up as experience grows. If you have a very stable operation, with good standard work, you may be able to move up to 95% of takt. Office operations are inherently more variable so a lower takt is appropriate – if you are able to use takt at all!

Then, it is necessary to adjust the times for changeover and for unnecessary operations. Player (Operation) 3 must wait 10 seconds when changing colours. Operation 4 is not done for Products 4, 5, and 7. Look at the data – the player card.

Operation 1: This time the game should be run with ONE PIECE FLOW. This means passing on one unit at a time instead of the batch. It will mean that each individual square needs to be cut out at operation 1 – not just cutting out the batch of products. So, the time taken to cut out one product square needs to be found.

Operation 3: A change in colour happens on average every second order. So add 10/2 = 5 seconds for this. The average order size is about 2.5. So then 5 / 2.5 = 2 seconds need to be added to the cycle time for Operation 4.

Operation 4: Half the products skip operation 4. So the observed time can be halved. Of course, it will be realised that these are rough calculations. But, as will be seen later, with kanban, CONWIP, and bucket brigade such approximations are generally adequate, and are self-correcting.

The table below shows an example. The times will, of course, depend on the dexterity of the people and on the thickness of the pens. The astute will recognize that, since the shapes are cut out later on, it is not important to do very accurate colouring in on the earlier operations.

Here the line is balanced against a planned cycle time of 12 seconds by accumulating the individual operation times. In your game, you may well get another answer!

| Station | Time | Adjusted time | Balance |
|---|---|---|---|
| 1 | 9 | 9 | 1 operator (cut one) |
| 2 | 5 | 5 | 1 operator |
| 3 | 11 | 11 + 2 = 13 | 1 operator (tight) |
| 4 | 8 | 8 / 2 = 4 | 1 operator for both 4 and 5 |
| 5 | 4 | 4 | |
| 6 | 23 | 23 (2x12) | 2 operators |
| 7 | 15 | 15 (2 x 8) | 2 operators |
| 8 | 10 | 10 | 1 operator |

So here we need a net of 1 extra operator (player) if we are to meet the planned cycle time of 12 seconds. Notice that operation 3 slightly exceeds the planned cycle time. But we have been generous in our 80% assumption. Of course, the combined times must all be less than 12 seconds – except station 3. Operations 4 and 5 are carried out as one operation, without interruption.

There is another question: can the operator at station 2 do any other task while waiting for the changeover – assuming that the process does not need his participation?

And another question: can operators do self checks on quality in their unused time? If the two operators at station 7 can take over inspection, then we are back at the same number of operators as used in the first cut and fill game.

You may make use of the Operator Balance Board (also known as 'Yamazumi' Board) on a following page. Draw a horizontal line for the takt time. Draw a second horizontal line for the Planned Cycle Time. Then accumulate the activity times for each Operator making sure that no cumulative time exceeds the Planned Cycle time (or maybe just exceeds it). You should end up with a stacked bar chart, also known as a 'Manhattan Diagram'. See the example slide from the Paperwork Office Cell Game.

Now set out the line to reflect the balance. Distribute copies of the slide and ask players to complete their own balance board.

Once the various player allocations have been decided, run the game as a push system over 10 minutes and record the scores. Again, prime the line with one piece of product 3 in front of each station.

But, this time, play the game with ONE PIECE FLOW. At station 1 the first operator cuts out individual squares (NOT SHAPES). Product squares are passed along one at a time, rather than in a batch.

Is it possible to identify the bottleneck(s)? By accumulation of inventory? If the line was well balanced AND IF THERE WAS NO VARIATION IN PRODUCT OR PROCESS there should be little accumulation. Was this the case?

Get the group to comment on the success or otherwise of the exercise.

Note: In this and other games following, as an option, consider using a U shape cell rather than a line.

In this case operators can jump stages, working on both legs of the cell. In particular, consider whether the first or second operator can also do the final task. If so, output can be regulated.

# Cut and Fill Game 3: Line Balance with Kanban Pull and One Piece Flow

A Kanban pull system can stabilise inventory and expose problems. But it is not without a risk of a drop in performance with mixed model products.

Run the game again using the balanced line developed in the last game, but now insert a kanban square between each operation.

The rule is, if the square is full, stop work at the station feeding the square.

In the case of branching operations, such as in the last game, two kanbans are set out – one in front of each parallel station. If either square is empty, the previous operation is authorised to fill it. If both squares are occupied, the previous operation stops.

Note: In the example game, there is no kanban square between operations 4 and 5.

For the sake of comparison, prime the line with one piece of inventory on each square. This then allows pure pull. Pure pull happens when a customer pulls a product from the end of the line. If the game is started with zero inventory it will be a push system that eventually develops into pull.

Check the players during the game. There is a natural tendency to want to push, thereby ignoring the kanban square.

It is possible that there will be a drop in performance compared with the previous game. Why? Because a player has to look ahead as well as wait for available products to work on. Thus there is a double requirement or constraint.

Get the group to comment on the success or otherwise of the exercise. Bottlenecks are now more apparent. Non-bottlenecks will stop working from time to time but bottlenecks will be very busy. Although now running much better than in earlier games, there is still a problem with different product work content. Cutting out a batch of arrows is a particular problem, causing surges of work. Two thirds into the game there is a lot of cutting out involved. This leads into the next games.

# Cut and Fill Game 3A: Line Balance with Kanban Pull and One Piece Flow, Mixed Model and FIFO Lanes. Possible Heijunka?

A modification to the last game involves running 'mixed model' production. Here, long tasks, such as cutting out the arrow, are followed by short tasks such as cutting out the triangle. This game is deliberately complicated by the fact that there is a changeover operation at step 4.

An answer is mixed model: long tasks, mixed with short tasks and small batches.
Ask the group to devise a mixed model sequence for the first 40 products. There are many alternatives.

First, the group will need to decide on a reasonable batch size to cope with the changeover. If there is no changeover, pure mixed model with one-piece flow is the best solution.

Let us say that a batch of two or three is decided on. This needs to be checked against feasibility – that it is possible to do that many changeovers. Then, from the schedule, accumulate the products. This is shown in the table below.

| Product | Batches | Total |
|---------|---------|-------|
| 3 | 2, 4 | 6 |
| 4 | 4, 3, 3, 5 | 15 |
| 5 | 1, 2, 2 | 5 |
| 6 | 4, 3 | 7 |
| 7 | 3, 2 | 5 |
| 8 | 1, 3 | 4 |
| | Total | 42 |

Now using a batch size of two or three, develop a mixed model sequence. A possibility is shown below.

| Period | 1 | 2 | 3 | 4 | 5 | 6 | 7 | 8 | 9 | 10 |
|---------|---|---|---|---|---|---|---|---|---|----|
| Product | 7 | 3 | 5 | 4 | 6 | 7 | 4 | 3 | 5 | 4 |
| Quantity | 3 | 2 | 2 | 3 | 3 | 2 | 2 | 2 | 3 | 3 |

| Period | 11 | 12 | 13 | 14 | 15 | 16 | 17 | 18 | 19 | 20 |
|---------|----|----|----|----|----|----|----|----|----|----|
| Product | 6 | 8 | 4 | 3 | 5 | 4 | 6 | 4 | | |
| Quantity | 2 | 2 | 2 | 2 | 0 | 2 | 2 | 3 | | |

A new schedule reflecting the sequence will have to be given to the first two stages.

We are now moving towards Heijunka – leveling the schedule across the day. A real Heijunka sequence would expect target output at a 'pitch' timing – where a pitch time is the batch x cycle time. So 3 products are supposed to be completed in 3 x 12 = 36 seconds, then the next 2 in 24 seconds, and so on.

Once again, the game is run with kanban squares between stations. However, a FIFO Lane (First In First Out) can be established where there are expected to be accumulations. For instance, in front of Step 4 during changeover, and in front of Stage 6 during longer cut out times. A FIFO lane is just an extended kanban square, where products move in FIFO sequence. So, instead of a kanban square of one in front of these stations, there is a FIFO lane of (say) 3. Kanban rules apply. No more than 3 are allowed to accumulate or the previous operation stops work.

Set this up, with mixed model and FIFO. Prime the line. Layout and stations as with the Line Balanced game. Run and discuss. Complete the Measures sheet.

# Cut and Fill Game 4: Line Balance with CONWIP

CONWIP, for constant work in progress, is a simple but highly effective method for this game. It combines push and pull.

Simply, as one product is completed at the end of the line, another is started at the beginning of the line. Hence the total inventory in the line remains constant.

No kanban is used. Players simply push products onto the next stage as they complete their work.

CONWIP is an effective way to run a line without complicated timing and line balancing. The longest operation will determine throughput. With CONWIP inventory accumulates naturally where it is required – in front of the longest operation.

Set this up. Use the initial Order Schedule or the Mixed Model Schedule developed in the last game. Prime the line. Layout and stations as with the Line Balanced game. Run and discuss. Complete the Measures sheet.

# Cut and Fill Game 5: Bucket Brigade Balancing

Bucket Brigade balancing is a heuristic method that self-balances the line. Bucket brigade is potentially more efficient here because of the varying work content between products. This is a major advantage.

Bucket Brigade Balancing is an effective way to balance any line without having to time operations. It 'self-balances'.

Bucket Brigade line balancing is fully described in the Airplane Bucket Brigade game. Please read those instructions as to how bucket brigade works.

In this game, unlike previous cut and fill games, all operators will need to stand and walk. They cannot sit, because they need to move between stations.

This game is a more complex case because there are stations that exceed the required takt time.

One solution is to simply to run two independent parallel lines, each with any number of operators, up to four. The schedule is shared between the two lines. As with Airplane Bucket Brigade, you begin by selecting the required number of operators and then allow them to self-balance. If demand falls, reduce the numbers of operators. If demand increases, increase the number of operators.

Set up the game by laying out the stations. For instructions on how to self-balance, read the Airplane Bucket Brigade game instructions.

Note that, as with the earlier Bucket Brigade Game, two rounds can be run. The first uses traditional line balancing, the second uses the bucket brigade method. The second round will produce higher throughput at lower WIP inventory. Please refer to the Airplane Bucket Brigade results sheet.

## Summary of Learning Points

- Operator cycle times must be established for traditional line balancing. But with multi products and multi stations this becomes unfeasible. So take the average product at each station and make adjustments. Exact precision is not required because of variation and the learning effect.
- Improve on timing by experience. Monitor the line for bottlenecks and the learning curve effect.
- Calculate takt and planned cycle time as a percentage of takt time.
- Balance against planned cycle time, not takt.
- With experience and improvement, increase the planned cycle time percentage. 95% may be maximum possible because of variation.
- Operations in parallel is a way to balance the line where operation cycle time exceeds the planned cycle time.
- Kanban squares stabilise inventory and help to highlight problems.
- But, kanban can decrease throughput.
- FIFO lanes are preferable instead of kanban squares, in front of operations with changeover or complex product operations.

- CONWIP involves both push and pull.
- CONWIP is an effective way to run a line without complicated timing and line balancing. The longest operation will determine throughput. With CONWIP inventory accumulates naturally where it is required – in front of the longest operation.
- Bucket Brigade Balancing is an effective way to balance any line without having to time operations.

**Kata**

Here the 5 step Kata method can be used as an option. Get the players to articulate the target condition, select the change, run and compare against predictions, and record the learning experiences.

**References and Further Reading**

- John Bicheno and Matthias Holweg, *The Lean Toolbox*, 4th edition, PICSIE, 2009, Chapters on Cell Design and Layout
- Mike Rother, *Toyota Kata*, McGraw Hill, 2009, Part III

# 5S – Numbers

## Key Areas

5S, Waste, Housekeeping, Productivity, Variation reduction, Standards

## Manufacturing or Service?

Both. 5S is a basic Lean tool for variation reduction, the establishment of consistent standards, and visual management. It is also about safety. And it is about housekeeping. Note the order of these. Unfortunately some organisations have emphasised housekeeping to an excessive extent. This has given the impression that Lean = 5S. Of course, that is not only wrong but trivialises Lean.

## Overview

This 'classic' Lean game or exercise is played individually by any number of participants. It is played over a number of rounds. In each round, productivity increases, generally variation is reduced, and errors or omissions are easier to detect.

Note the other 5S game as part of the 'Squares Game' Series. This is a more complex game, taking a little longer, useful for 5S in a team setting.

## Summary of Learning Points

- When things are in a mess productivity is affected and variation is greater
- How much time is lost looking for things?
- The 5S's are Sort, Simplify, Scan, Standardise, Sustain. There are variations on this; for example, Simplify may be replaced by Straighten; and Scan by Shine.

## Players

Any number

## Approximate Time

20 minutes.

## Description

There are four sheets, each with numbers 1 to 64. A sheet is given to each player, face down. They are told that the standard time is one second per number.
Upon starting, players must circle each consecutive number. Time 65 seconds, then call 'stop'. Write up the data. Repeat with the next sheet, and so on. Discuss the implications.

**Instructions**

1. Prepare sufficient copies of each of the four sheets (4 x number of players). Distribute Sheet 1 to all players, face down. Ask players not to turn over.

2. Explain that this is an exercise to illustrate the concept of '5S'. Do not explain what 5S is at this stage. The task is to circle consecutive numbers. No numbers may be skipped. The standard time is 1 second per number. You will have 64 seconds. Number 1 is to be found in the top left hand corner.

3. Turn over and begin. Time 64 seconds. Call stop.

4. On a flip chart record how far each has got. Write down the number reached by each player. Point out the wide divergence between the best and the worst.

5. Now say that there are five 'S's in a 5S programme. These are Sort, Simplify, Scan, Standardise, Sustain. Now we will do the first S, sorting. It also includes basic tidy up. That means throwing out all excess, unnecessary materials – but in this case numbers. The numbers will also be re-oriented. This has been done for you in Sheet 2.

6. Hand out Sheet 2, face down. When turned over, it will be the same task – circling the numbers.

7. Turn over and begin. Time 64 seconds. Call stop.

8. On a flip chart record how far each has got. Write down the number reached by each player. Point out that progress has been made.

9. The next S is Simplify. "A place for everything and everything in its place'.

10. Ask if anyone noticed a pattern of the numbers. Some will point out that, with few exceptions, the numbers occur in a sequence of columns, then rows.

11. Give out Sheet 3, face down. Say that rows and columns have already been added.

12. Say that, this time, players should call out when they have circled all 64.

13. Begin. Observe your watch, and record the times in seconds when players call out. When using a watch, start this round at '12' or '6' on your watch to make recording easy.

14. Say that there is still room for improvement. Some numbers are misplaced.

15. Now say that the next S, Scan, is about continual maintenance. Putting back anything that is out of place. This S will not be demonstrated.

16. The next S is Standardise. In the numbers exercise there will be a specific slot for each number.

17. Give out Sheet 4, face down. Same task. Same call out when reach 64. But also note any anomalies such as missing figures. If they are sure that a number is not there, it may be skipped.

18. Turn over and begin. Observe your watch, and record the times in seconds when players call out.

19. Point out that this was not only much quicker, but players were able to pick up errors and omissions at a glance.

20. Now go to the flip chart with the four sets of figures. Say that 5 S is principally about 3 things, plus two others:
    - Productivity: notice the great reduction, steadily, from Sheet 1 to Sheet 4. It is almost everyone's experience that they work better in a tidy, well organised situation.
    - Variation: notice the reduction in variation, as shown by the spread of results, over the sheets
    - Problem identification: with Sheet 1 it was very difficult to notice if a number was missing, misplaced, or the wrong size. By Sheet 4, this was virtually instantaneous.
    - Safety: the exercise did not illustrate safety, but a tidy environment is likely to be safer. Some organisations call their program '6S' to incorporate safety.

- Housekeeping: this can be important in messy environments. It can also save space. But the first four points above are more important.

**Forms, Graphs, Tables, Typical Results**

Forms 1 to 4 are reproduced. Make copies for players.

Typically, players get into the mid 20's with Form 1, the mid 40's with Form 2 – some may even complete. Form 3 is finished in around 30 seconds, form 4 in 5 to 10 seconds.

**Instructor notes and discussion**

The forms are deliberately not perfect:
- 50 and 51 are interchanged
- 3 and 11 are interchanged
- the font of 19 is smaller than expected
- the font of 6 is bigger than expected

**Equipment**

Make sufficient copies of Forms 1 to 4.
Flip chart and pens.

**References and Further Reading**

Bicheno and Holweg, *The Lean Toolbox*, 4th edition PICSIE Books, 2009, pages 78 to 82

# Cell Design and Balancing

**Key Areas**

Cell design, layout, takt time, line balance, Standard operations chart, waste removal

**Manufacturing or Service?**

Manufacturing

**Overview**

This is a detailed analysis of a small manufacturing cell – showing the 'before' activities in detail. Someone else has prepared the process activity chart, and now players must analyse and improve the cell.

Players are arranged into team of 3 to 5. They are presented with process activity data, current demand, and current layout. The task is to determine the current number of workers required, via the takt time. Then the activities and layout are critically assessed to run the process with fewer workers. This involves a line balance. Optionally, a standard operations chart can be drawn.

**Summary of Learning Points**

- Calculate takt time
- Decide an effective planned cycle time, less than takt
- Calculate number of workers
- Ignore machine time
- Improve layout
- Use the Operator Balance Chart
- Ergonomic considerations
- Balance the line
- Draw the operations chart

**Players**

Any number up to about 20, in groups of 3 to 5. More players with less coaching.

**Approximate Time**

1 to 1.5 hours

**Description**

Players are given sheets of activities and layout. They must analyse, discuss, and come up with a new layout that reduces waste, reduces the number of workers, and improves flow.

**Player Instructions**

- Each member of your team has been given three sheets – activity and demand data, the existing layout, and an operator balance chart.
- Calculate takt time.
- Use 90% of takt time to calculate the current number of workers. You may use the Operator Balance Chart. Your instructor will tell you the formula if you cannot figure it out. (Beware, there are traps built in!)
- Examine the activities and current layout. Identify value added activities and waste.
- Redesign the layout. You will have to make several reasonable assumptions. Think creatively. There are several good alternatives. There is no one best way.
- Draw the new operator balance chart.
- Be prepared for your team to present your solution.

**Forms, Graphs, Tables, Typical Results**

See accompanying diagrams

**Instructor notes and discussion**

- Hand out the sheets
- Takt time = Available time / Demand = (480 – 30 – 40 – 10)/800 = 0.5 mins = 30 seconds
- Use 90% of this, or 27 seconds. If 100% is used, variation will ensure that the target is missed. 90% is a matter of judgment, not a rule.
- By the way, why 90% of takt time? Because of variation! Re-visit the Muda, Muri, Mura game. Here you have learned that 100% loaded (or 100% 'utilisation') leads to very long queues and delays. Toyota refers to this as 'Planned Cycle time'; Hewlett Packard has used the term 'under capacity scheduling'.
- Number of workers = (Sum of human activity times) / 27
- Alternatively the operator balance chart can be used. Draw the takt time line and cycle time line, and accumulate times in columns until the cycle time line is reached. Then start a new column.
- Note that machine times must be excluded – if they are included you are saying that the operator stands and watches the welder.
- Decide on value adding and non-value adding activities. Note that there are very few value adding activities – only welding and assembly – total 28 seconds – and of this time 18 seconds is machine time.
- Note that proportional times must be calculated. Thus, for example 120 / 30 for the fetch raw material time. Similarly for de-burr.
- The layout can be considerably improved. A good solution is a U shaped cell arrangement. Limited walking is fine. Sitting is not desirable for health and flexibility reasons.
- Avoid reaching down into stillages, and twisting if standing in one place.
- Walk times must be estimated.
- You should be able to run the new cell with two operators.
- De-burr is waste. This may not be able to be eliminated in the short term, but should be the target for elimination by kaizen over the medium term.
- Generally, a good cost-effective improvement is to have a 'quick release' device that kicks out the part after welding. This will save a few seconds.

- In doing these exercises, the team should target getting to two operators, first. Although it is possible to go below this, will further reduction lead to cost saving? Not unless the part-used operator can be used elsewhere.
- It is generally not good practice for operators to leave the cell to go and fetch raw material. This should be done by a material handler/runner. A Kanban system should be arranged.
- Give consideration to alternating the work (say) once per hour. To do a 30 second repetitive cycle all day is very tough.

## Equipment

Paper and pencil exercise. Flip charts to present solutions. Also see the tables on the following pages.

## References and Further Reading

Bicheno and Holweg, *The Lean Toolbox*, 4th edition, PICSIE, 2009; Chapter 8

Mike Rother, *Toyota Kata*, McGraw Hill, 2010, Part 3, and Appendix 2.

## Manufacturing Cell Layout and Balance

The demand from a process is 800 products per day. The area works an 8 hour day, with a lunch break of 30 minutes, two 20 minute refreshment breaks, and an additional 10 minutes allowed for comfort.

The part to be made comprises a welded part and a bush. The parts weigh 2.5 kg.

The activities performed are given in the following table.

| Activity | Description | Time |
|----------|-------------|------|
| A | Fetch inbound material from raw material stillage (every 30th cycle) | 120 |
| B | Walk to inbound raw material container | 5 |
| C | Pick up raw material | 2 |
| D | Walk to left side of weld machine | 3 |
| E | Place raw material on left of weld machine | 2 |
| F | Walk to centre of weld machine | 2 |
| G | Remove welded part from jig on weld machine | 5 |
| H | Place welded part in outbound area | 2 |
| I | Walk to left side of weld machine | 3 |
| J | Pick up raw material from left side of weld machine | 3 |
| K | Move raw material to centre of weld machine | 2 |
| L | Secure raw material into jig | 6 |
| M | Start welder | 2 |
| N | Welder runs through cycle | 18 |
| P | Wait for part to cool | 15 |
| Q | Pick up welded part and jig from weld machine | 2 |
| R | Visual inspection of part | 6 |
| S | Carry part to bush area (3 out of 4 parts on average) | 3 |
| T | Carry part to de-burr area (every 4th part on average) | 4 |
| U | De-burr (every 4th part on average) | 12 |
| V | Carry de-burred part to bush area (every 4th part) | 2 |
| W | Fetch bushes container (every 200 cycles approx) | 200 |
| X | Pick up bush | 2 |
| Y | Walk to assembly table | 3 |
| Z | Assemble product | 10 |
| A1 | Place assembled product in finished goods (FGI) area | 3 |
| | | |

# The Paperwork Cell (Office Kaizen)

## Description

Players are given sheets of activities and a layout. They must analyse, discuss, and come up with a new layout that reduces waste, reduces the number of workers, and improves flow.

## Key Areas

Cell design, layout, takt time, line balance, Standard operations chart, waste removal – but in a warehouse or dispatch bay context. Could be adapted for invoicing or other transactional processes.

## Manufacturing or Service?

Service

## Overview

This is an exercise based on an actual dispatch bay in an office products warehouse, where invoices are added to picked boxes of items.
Players are arranged into teams of 3 to 5. They are presented with process activity data, current demand, and current layout. The task is to determine the current number of workers required, via the takt time. Then the activities and layout are critically assessed to run the process with fewer workers. This involves a line balance. Optionally, a standard operations chart can be drawn.

## Summary of Learning Points

- Calculate takt time.
- Decide an effective planned cycle time, less than takt. Why?
- Discuss the relevance of takt in an office environment.
- Calculate number of workers from manual times and planned cycle time.
- Ignore machine (printer) time. If you include this, the person is standing and watching the printer.
- Improve the layout.
- Use the Operator Balance Chart.
- Ergonomic considerations. Is the counter ergonomic?
- Balance the work? Is it better to have operators equally loaded, or one loaded up to planned cycle, while the other has some slack time?
- Draw the operations chart.

## Players

Any number up to about 20, in groups of 3 to 5; more players with less coaching.

**Approximate Time**

1 to 1.5 hours

**Player Instructions**

- Calculate takt time (based on anticipated figures).
- Work out a suitable Planned cycle time.
- Calculate the required number of office workers for one-piece flow. Make allowance for additional activities.
- Go through the list of activities and identify the 'value adding' steps.
- Decide on an appropriate target number of office workers.
- Examine the activities and layout critically, and make improvements. Show the new layout on a Layout Diagram.
- Show the revised way of working on a Work Balance Board.

What is allowed and what is not:

- Anything can be changed or rearranged, but
- No equipment can be placed on the counter for security and safety reasons
- A paper invoice is still required. You cannot simply assume that the process can be computerised or that invoices can be sent by e mail. It is standard current practice to include paper invoices, showing all the items picked, in with the parcel.

**Forms, Graphs, Tables, Typical Results**

Activity sheet given below. Operator Balance sheet as for Manufacturing Cell Game (i.e. the previous game).

**Equipment**

Player sheets: use Activity List Sheet below, and Operator Balance Sheet from Manufacturing Cell Game.

**Instructor notes and discussion**

- Hand out the sheets
- Takt time = Available time / Demand = (8*60 − 30 − 20) / (760+100) = 430/860 = 0.5 mins = 30 seconds
- Use 80% of this, or 24 seconds. If 100% is used, variation will ensure that the target is missed. 80% is a matter of judgment, not a rule.
- By the way, why 80% of takt time? Because of variation! Re-visit the Muda, Muri, Mura game. Here you have learned that 100% loaded (or 100% 'utilisation') leads to very long queues and delays. Toyota refers to this as 'Planned Cycle time'; Hewlett Packard has used the term 'under capacity scheduling'. In a factory, that has less variation one can use 90% or even 95%, but most paperwork / office operations have higher variation, so require a greater allowance.
- Number of workers = (Sum of human activity times) / 24. Here, this is not as straight forward as just adding up the times. Players will have to calculate a weighted average.

Note that proportional times must be calculated. Thus, for example, entering changes and reasons that occur 20% of the time, and walking for the batch.

- Alternatively the operator balance chart can be used. Draw the takt time line and planned cycle time line, and accumulate times in columns until the cycle time line is reached. Then start a new column.
- Note that printer times must be excluded – if they are included you are saying that the operator stands and watches the printer.
- Decide on value adding and non-value adding activities. Note that there are very few value adding activities – only printing out the one copy.
- Several activities are necessary non-value adding.
- Some activities could be open to debate: is an envelope necessary? Stamping 'checked'?
- Is one-piece flow preferable here? It probably depends on shipment frequency and packing arrangements.
- The layout can be considerably improved. A good solution is a U shaped cell arrangement or desks arranged along a table next to the counter. Limited walking is fine. Sitting is not desirable for health and flexibility reasons.
- Walk times must be estimated.
- You should be able to run the new cell easily with two operators. One is possible but only with considerable modifications. For example can a pokayoke or self-check be done by the pickers?
- In doing these exercises, the team should target getting to two operators, first. Although it is possible to go below this, will further reduction lead to cost saving? Not unless the part-used operator can be used elsewhere.
- It is generally not good practice for operators to leave the area to go and deliver forms. This should be done by a material handler/runner. A Kanban system could be arranged.
- Give consideration to alternating the work (say) once per hour. To do a 30 second repetitive cycle all day is very tough.

### References and Further Reading

- Bicheno and Holweg, *The Lean Toolbox*, 4th edition, PICSIE, 2009; Chapter 8

- John Bicheno, *The Service Systems Toolbox*, PICSIE, 2012

### The Paperwork Cell

The Paperwork Cell is the last stage before final dispatch in an office products company that takes orders over the phone and internet, picks the orders, and ships overnight. Orders are received throughout the day. Average demand is around 760 orders per day, but is expected to rise by around 100 per day. The office works an 8 hour day with one 30 minute break and two 10-minute breaks. The following is a process activity map collected from video sequences. The times are median times in seconds for processing a single order. Assume that the workload is evenly distributed throughout the day. This is unrealistic, so you should build in an allowance for variation

**Activity List**

| Activity No | Unit Time (secs) | Activity |
|---|---|---|
| A | 3 | Walk to picklist box on counter |
| B | 2 | Pick up picklist |
| C | 3 | Reorient and sort picklist |
| D | 3 | Walk to desk |
| E | 1 | Place picklist in in-basket |
| F | 3 | Walk to computer chair and sit down |
| G | 2 | Pick up picklist from in-basket |
| H | 8 | Scan picklist bar code and wait for order to be retrieved on computer screen |
| I | 15 | Check picklist against original order. All items picked? (Y/N: if Y (80%) go to activity L |
| J | 24 | N: Enter changes to invoice on computer (20%). |
| K | 8 | Enter reason code for shortage into computer |
| L | 3 | Check paper, insert envelope and start printer |
| M | 15 | Print out 2 copies of the invoice, and an envelope |
| N | 3 | Separate office and customer copies of the invoice |
| O | 6 | Stamp 'checked', date and clerk number on customer copy |
| P | 4 | Place customer copy of invoice into envelope |
| Q | 2 | Place envelope into customer invoice tray |
| R | 2 | Place office copy into file tray |
| S | 8 | (Every approx 10 envelopes) Walk from customer tray to dispatch box, walk back |
| T | 15 | (Every approx 10 copies) Take office copy to filing cabinet and file, walk back |

# The Area Supervisor – E Mail exercise
*A Day in the Life (of a new Lean Manager)*

## Key Areas

Implementation, leader standard work, priority, span of control, delegation, system, flow

## Manufacturing or Service?

Both. Some words in the exercise may need to be adjusted for your own environment.

## Overview

This is what used to be called an 'in basket exercise' – now re-named e mail exercise. A supervisor, in charge of several cells, and relatively inexperienced in Lean is battling various conflicting demands. He or she must make decisions. But the real opportunity is to re-think an appropriate management style that will save the area from a Lean implementation that may otherwise be doomed.

## Summary of Learning Points

The list of emails is a vehicle to discuss Lean implementation. However, the root problem may be a lack of leader standard work: regular routine, managing at the Gemba and reviews taking place every day, with escalation when necessary.

There are no right answers. Some points that may arise include the following.

- Is the error trend increasing or is it natural variation? If the trend is increasing look at the process, not the person. What paretos exist?
- If Lean meetings are postponed, this sends out a negative message on importance. What to do?
- Off sick: this is the middle of the day! Why now? Are cell leaders multi-functional? What mechanisms exist for working together? Should the supervisor be involved? Visual management boards?
- TPM report. Why not presented or discussed? Maybe no problem.
- Visual management needs constant updating. Persons responsible? A daily task? But why only noticed now, rather than at morning meeting? Are checklists (especially operator checklists and activities) being put in place?
- Shortages: why inform the supervisor? All sorts of possibilities here. Start with a table of A, B, C parts against runners, repeaters, strangers to fix appropriate course of action.
- Variances: is variance tracking compatible with Lean? Variances may be completely out of the supervisor's control. Variances also may result from incorrect budgets. Beware of a variance witch-hunt with Lean. Are variances understood? Reducing the variance may, for example, lead to overproduction. This comment opens up the possibility of discussion on Lean Accounting.
- Bottleneck. Good that there is an apparent awareness of bottleneck. This opens up discussion on bottleneck priorities. For example, alternative machines, routings, TPM priorities, buffers. Drum Buffer Rope: are parts before the bottleneck still being built? Early warnings of bottleneck stoppage?

- Bottleneck 2: but, missing the morning meeting is a slippery slope. This may be the root cause.
- KPI: is this organisation KPI driven by 'remote control'? Get the MD to the Gemba!
- Design: involvement of supervisor sounds positive.
- Late: why from the MD?
- Shortages: see above. Knock on effects?
- Shortages 2: 'Ahead of the game'. Careful! Making ahead of requirements is a bad sign. How is such a situation authorised? Day-by-the-hour manufacturing?
- IT Software: of course, the software may make a difference, but only if good process analysis has been done first. 'Don't lock in waste'; 'Don't automate, obliterate'.
- See Lean meeting postponement. More evidence?
- Christmas Lunch: an opportunity to review, thank? Or just to help establish good working relations? Would this be the only time operators get to see the MD?
- 5S. See Lean commitment above. 5S needs to work through 'pull' or need, not through push, where it may be seen as optional.
- Policy Deployment. This could be the opportunity for bringing together the Lean transformation.
- Skeptical. Of course! See 5S above.

**Players**

Groups of 3 to 5, each discusses and presents back to the full group. Maximum 20?

**Approximate Time**

1 to 1.5 hours including discussion

**Player Instructions on next page**

**Email list**

You are an area supervisor, in charge of five team leaders from five cells, 1 to 5.  Yesterday you were out of the office, attending a course on motivation connected with the Lean implementation that has been in progress at your company during the past six months.  You have just arrived in the office. You normally arrive 30 minutes before work starts. Today is Wednesday. You open your emails and find the following sequence of messages. Cells 1 to 3 form a sequence. Cells 3 and 4 are specialised.

| Date and Time | From | Designation | Message |
|---|---|---|---|
| Mon  14:05 | Peter O | Quality Mgr. | The trend in errors is increasing. What actions are you taking? |
| Mon 14:09 | Gail | MD's Secretary | Lean Project meeting has been postponed to Wednesday 09:00 |
| Mon 15:20 | George P | Cell Leader 4 | One person just off sick. Another on leave until next Monday. Help? |
| Mon 16:00 | Peter W | Maint Mgr | TPM report now ready |
| Mon 16:55 | Peter O | Quality Mgr | Just noticed that the graphs on errors have not been updated since last Wed. |
| Mon 17:10 | Clive | Stores Mgr | Still shortages of cartridges. Seems to be getting worse, with the new supplier. |
| Mon 17:40 | John D | Accounts | Can we meet to discuss the unfavourable variances in your section? |
| Tues 07:10 | Mark | Cell Leader 1 | I will not be able to take a morning meeting today. The bottleneck op was down yesterday morning. |
| Tues 07:22 | Simon E | MD | I am still waiting for last week's KPIs!! |
| Tues 07:29 | Darrell | Design Mgr | Please would you bring your comments on the new design to the meeting this afternoon. |
| Tues 08:07 | Simon E | MD | Just had a call from Strimmers. The wrong number of W was delivered. Fix! |
| Tues 09:10 | Sally T | Cell Leader 3 | At this morning's meeting delivery shortages is still an issue. |
| Tues 09:16 | Nicola | Cell Leader 2 | Our backlog of is slowly increasing. We are short of cartridges, but are ahead of the game on F's |
| Tues 10:55 | Mike | IT Mgr | I have been contacted by BatchTech. They believe the latest release of the scheduling software will make a big difference. Can we discuss? |
| Tues 11:02 | Nick | Kaizen Mgr | Next week's kaizen event in Cell 1 seem doubtful. They are too busy! |
| Tues 11:20 | Margaret | Canteen Mgr | Reminder: This afternoon's meeting on the Christmas lunch |
| Tues 11:42 | Nick | Kaizen Mgr | We need to re-think 5S. It is not working. |
| Tues 14:30 | Simon E | MD | What are your thoughts on Policy Deployment? |
| Tues 15:00 | George P | Cell Leader 4 | Just back from the Lean training course. My guys are a bit sceptical. |

# Call Centre : Failure Demand and Targets

**Key Areas**

Call Centre, Service Centre, failure demand, abandoned calls, rework, capacity, queues

**Manufacturing or Service?**

Service but also found in manufacturing organizations, especially call centres
May also apply in Hospital A&E

**Overview**

This simulation enables managers to appreciate the consequences and trade-offs of setting target times for calls. Is there an optimal value for the number of servers (say in a call centre)?

Many call centres have process-based metrics. A few have outcome-based metrics. This simulation allows both to be experimented with.

Service centres can be counterintuitive because using averages but ignoring variation in arrivals patterns and process times is dangerous. ("Drowning in a river of average depth 1m.")

Some call centre and service centres set or impose target times. This is done for efficiency or productivity reasons. At the other extreme the on-line shoe US shoe retailer Zappos has the policy of no call length limitation and their service operators will stay on the line as long as required to satisfy the customer. The longest call has been 9 hours, apparently.

The game can also be used in a context of field service operations where visits are abandoned where they are too long or complex, resulting in the necessity for a repeat visit. In this simulation it will be assumed that if a call is abandoned because it is taking too long, the caller will call again.

John Seddon of Vanguard introduced the now well-known concept of failure demand – demand resulting from not doing something or not doing something correctly. Failure demand adds to the load or queue of work, and can overload the system. This is 'Muri' as explored in the Muda, Muri, Mura game.

The game makes use of the Beta-like (or Erlang-like) distribution of call lengths, which is typical of the vast majority of service times both for calls to a call centre and for field service calls. The game uses a Poisson distribution for arrivals that is also typical.

A 'feel' for call length distribution and arrival rate is interesting and highly relevant to all service managers.

The game can be set up as a competition between two call- or service-centres.

**Summary of Learning Points**

- As operator utilisation increases, queue length increases exponentially.
- Setting targets or processed-based metrics can have a severe impact on queues and call centre efficiency, particularly if the target or abandon signal is set too low.
- Having an outcome-based metric allows operators to stay on the line until the caller is satisfied. In the game, this is simulated by letting operators roll until a six is obtained, however long that takes. In this case there would be little or no failure demand. Such policies may facilitate operator autonomy and can improve operator engagement.
- Setting targets with respect to maximum length of call often results in increased demand, especially if the target is set too low.
- 'Failure Demand' can be a significant cause of load and customer wait (queue) time.
- There may be a trade-off between call centre (or service centre) capacity cost and the cost of failure demand. Failure demand costs include customer wait time plus the hard-to-estimate costs of customers who 'defect' because they are fed up. This is an optional extension to the standard game.

**Approximate Time**

75 minutes typical. Longer if more rounds and discussion on costs are included.
The game can be extended by a number of options (see under **Extensions**, below). These can extend the game by an additional hour.

**Description**

- Smaller teams should play with 1 arrival (average) per period, bigger teams with 2 arrivals (average) per period. 2 is preferable. A big group could play with 3 arrivals per period.
- The game should be played over at least two rounds. Three rounds is preferable. Each round involves 40 periods of 20 seconds (13 minutes). One round should take place with no target time. That is, there will be no abandoned calls.
- Begin by distributing the call centre game flow sheet. The instructor should explain that call length depends on dice rolls – the number of rolls required to roll a 6. Ask how many servers are required. Write answers down for each team.
- The instructor should then do a demonstration of call lengths. Get one player to roll a dice. A call is completed if a 6 is rolled; it is not completed if a 1 to 5 is rolled. The player should simulate 20 customer calls. For each call record the number of dice rolls before a 6 is rolled. Move onto the next customer call when a 6 is rolled. This will take about 5 minutes for 20 calls.
- The instructor should draw the distribution on a chart. See two typical results. Explain that this is a Beta distribution typical of service centre call times. Ask whether this accords with players' experience. Say that this is the distribution that will be used in the game.
- Calculate the average rolls to get a 6. The theoretical average is about 5.5 rolls to obtain a six. Point out that this means that for an average of one arrival per period, at least 6 operators are required if there is one roll per period, and at least 3 operators are required if there are two rolls per period. BUT ONLY IF THERE IS NO FAILURE DEMAND. The game is played with two rolls per period.
- If players have played the Queuing Dice Game (Muda, Muri, Mura), remind them of this.
- In any case it may be a good idea to show players the graph of Queue time, variation, and utilization from the Muda, Muri, Mura dice game, after the game has been played

- The Instructor should demonstrate the game to the players by 'walking through' two or three customer calls. In doing so, say that the target has been set at three periods before abandonment.
- Tell the teams that there will be 2 (or 3 for larger teams) calls per period.
- Ask each team to set a team size and a limit on the number of rolls before abandonment.
- Ask the team to predict the queue length and average number of periods to complete a call. This is a form of Kata exercise.
- The instructor should select one of the tables from the tables of Poisson arrivals. The instructor should use the same table for each team. Select another table for each additional round.
- Start with zero queue.

## Players

- It is preferable to have two or more competitive teams. A game requires a minimum of either 9 or 12 people (7 or 10 of whom are operators), so the instructor should decide the number of parallel games.
- It is possible to play the game with one team of 8 people. This is the minimum. Here one players should assume the role of the call generator as well as handling the abandoned calls.

## Forms, Graphs, Tables, Typical Results

See Game slides

## Measures and Calculations

- After 40 periods, for each team, record or calculate:
  - The average number of periods to complete a call (for completed calls). To calculate this, it will be necessary to go through the completd customer call cards.
  - The number of abandoned calls. This number can be obtained by going through all customer call slips and adding the 'times abandoned' numbers.
  - The longest wait for completed calls. This number can be obtained by going through all completed customer call slips.
- Which was the better team? How accurate were your predictions?

## Instructions for Call Generator Players

- One player from each team is the call generator. Each 20 second period, that player should
  - Obtain the number of arrivals from the instructor. The number of customer arrivals is given in the Poisson table.
  - The instructor will call out the next period, every 20 seconds.
  - Each period, look up the number of customer arrivals from the Poisson table
  - Give each customer a number by incrementing the arrivals.
  - For each new customer, write the customer number and arrival period on a customer call slip.
  - Move the slips into the queue, adding the slips to the bottom of the queue pile.
  - This should be done at the beginning of the period, so that the slip can be started on by any available operator during that period.
  - Count and record number of slips in the queue before signalling that the next period may begin.
  - Check that all operators have completed all steps for the period.
  - Announce that the next period can begin.

## Instructions for Operators

- For the first period only, each operator takes a customer slip and rolls the dice once. For each operator: if a 6 is rolled, write the period number (1 in this case) on the slip and place the slip in the 'work complete' area. If 1 to 5 is rolled, the customer is carried over to period 2. No new slip will be taken. Add one to the rolls per customer record on the slip. (Use a 'gate' type record.). In period 2 a new slip will be taken.
- Note that each operator can only deal with one customer at a time.
- The instructor will call out the next period, every 20 seconds.
- Each period, each operator should go through the flow chart of Operator Instructions.

## Abandonment

There are two types of abandonment:

- Company policy abandonment, where the call centre management sets an upper limit on the time that a call should take. This is simulated directly in the game by fixing a limit. Setting this limit or target too low will generate large failure demand. Having no limit eliminates failure demand, but some calls may be excessively long, thereby affecting capacity and wait times for other customers. Is there an optimum?
- Customer abandonment, where a customer simply gives up waiting. This is not directly simulated, but can be incorporated by getting the Call Generator player to check the length of queue before every period. A maximum queue length can be set.

It is recommended that customer abandonment is not incorporated in a first round because, of course, average wait times will be affected. To know and experience the distribution of customer wait times is a valuable lesson.

## Instructor notes and discussion

- This is a powerful game, but needs practice and thought. Some instructors find aspects counter-intuitive even after playing the game.
- The instructor needs to arrange a display of the current period, either using a PowerPoint projector display or a flipchart.
- Operators are allowed 2 dice rolls per period to see if a 6 is obtained. (Rolling a 6 would indicate that the call has been completed.) This is because the average number of rolls to obtain a 6 is approximately 6. So an initial guess is that with 2 rolls per period it would take an average of 3 periods to service a customer. See below.
- There are two distributions of relevance in many service situations – the arrival distribution (very often Poisson), and the service distribution (Beta or Exponential is very common). This game gives players and managers experience in both. For service times, of course, the average service time is the time taken for servicing whilst servicing work is going on.
- The system being simulated is a single queue, multi-server situation. This is realistic, but note that there are other possibilities that are not considered here – such as priorities and reserved queues.
- If there is an average of one arrival per period, and the service time depends on the time to roll a 6, then it may be 'reasonable' to assume that the average service time is 6 periods. Hence, with two rolls per player per period it may be expected that 3 operators are required. Big mistake! Try it with Poisson arrivals of average one per period (see tables) and three operators. Queues build up. Why? Three reasons: Because unused capacity is lost forever, arrival variation is non uniform, and failure demand.
- But, hey, Lean is 'common sense', right?
- The game simulation may be surprising for players because:
  - Averages are dangerous in service situations. If the number of required operators is calculated on an average, managers will be in for an unpleasant surprise.
  - An aggressive target (i.e. a low number of rolls before abandonment) can be a silly thing to do. Such targets (maybe in the form of a time limit) are quite often used in call centres to 'motivate' operators. Of course, failure demand often results.
  - There is a non-linear relationship between utilisation and queue or completion time.
  - Failure demand can result in complete overload of the system.
- You should run a round with no abandon target limit and another with a similar average demand and number of operators, but with abandon at 6 rolls. There will be a fair increase in average queues, and a large increase in the maximum time to complete a customer demand.
- Allowing operators to stay on the line until the customer is satisfied is simulated by allowing operators to continue with a call until a 6 is rolled, however long that may take. There is no abandon limit or target. This means that the metrics would change from process-based to outcome-based. Employee engagement and employee autonomy are requirements. This can lead to an interesting discussion!
- Discuss if a time (or abandon) limit should be used. Compare the round where there is no abandon target with rounds where there is an abandon target.
- Importantly, in practice, 'abandoned' calls may not only result from the length of time to deal with the call request, but may also result from hurried and, unsatisfactory work that does not fully satisfy the customer. This is an important point that the instructor should bring out when explaining the game.
- Utilisation can be calculated as follows:

The Lean Games and Simulations Book

- o Average load / average capacity
- o Average load = average calls per period
- o Average capacity = (no of operators x rolls per period / average rolls for a 6). Since the number of rolls per operator per period is 2 and the average number of rolls to get a 6 is 6, then
- o Average capacity = (no of operators / 3)
- o And for 2 calls average per period, utilisation = 2 x 3 /(number of operators) (approximately) from  6 x 100 / (number of operators)
- It is instructive to draw the curve of average queue vs. utilisation. This is similar to the curve drawn in the 'Muda Muri Mura' Dice Game. Absolutely fundamental in Lean service!
- A discussion should be held between the teams as to which team was best, in terms of
  - o Customer satisfaction
  - o Overall system cost

## Extensions

1. An extension, and certainly a useful discussion, is to play a round, or several rounds, where abandoned calls do not go into the 'try again' queue but simply disappear. You should compare each of such rounds with the corresponding game(s) where abandoned calls do go into the 'try again' queue. Calls that disappear can make a big difference to the queue length and wait times – the shorter the 'abandon' target the bigger the difference between 'try again' and corresponding 'disappear' rounds. Here are the very important points:
   - Does the call centre know the proportion of abandoned calls that are lost forever, resulting in a customer defection?
   - Does the call centre know the proportion of incoming calls that are 'failure demand'? (That is, an apparently new call that is actually a repeat call resulting from abandonment.) Perhaps the only way to determine this is actually to listen in to what customers say.
2. A related extension is to play one or more rounds where, when the queue reaches a certain number, all new arrival calls are considered 'lost' and the call cards are placed in a separate pile. All new calls are placed in this pile until the queue reduces to a specified length (Try 8?) whereupon new calls go into the queue once again. This would repeat if the queue builds again. Here are the important points:
   - Does the call centre monitor the number of customers who call but give up before speaking to an operator?
   - What is the customer tolerance threshold for waiting?
3. As another extension, teams can calculate total cost for each round, assuming:
   - Operator cost £100 per round each
   - Customer delays cost £0, £2-50, £10 per customer per period. Of course, many service situations assume that customer delays are zero cost! They are not!
4. Finally, you can alter the arrival rate to simulate times of peak demand. If this is done, the round length should be considerably extended.

**Kata**

- Get the teams to predict performance
- Play a round
- Get the teams to agree on a target condition. "What is preventing us from ….."
- Make a change, and predict
- Play, measure, reflect

**Equipment**

- Slides
- Copy the Instructions for Call Cenerator Players and Operators. Players need to be clear on what they have to do and this may take some practice.
- Cut out Customer Call cards: make 80 cards per team for each round.
- One dice per player
- Two boxes per team (one for completed cards, one abandoned cards)
- Period display (best via PowerPoint and data projector)

**References and Further Reading**

- John Bicheno, *The Service Systems Toolbox*, PICSIE, 2012 – especially sections on 'Muda, Muri, Mura', variation, queues, failure demand.
- See the Muda Muri Mura Game and discussion points earlier in this book.
- John Seddon, *Freedom from Command and Control*, Vanguard, 2003
- Wallace Hopp, *Supply Chain Science*, Waveland, 2008, Chapter on Variation
- Any book on Queuing Theory. In USA queuing theory is known as Waiting Line theory.

# TPM Check List
*TPM: Developing a list of checks for your Car using Five Senses*

## Key Areas

TPM, Gemba, vision, checks, inspection, operator centered maintenance, autonomous maintenance, FMEA

## Manufacturing or Service?

Principally manufacturing, but applications in some service organisations

## Overview

This is a 'brainstorming' exercise that asks participants to develop a list of checks that can be done on a car using 'your God-given senses' (Peter Willmott). The list assumes no expertise as a car mechanic, but, if carried out, would certainly improve the safety, life, reliability, and probably comfort of your car.

This game is used by UK TPM Doyen, Peter Willmott. Thanks to him for the idea.

You can move on from a car to an everyday product such as a toaster or coffeemaker.

## Summary of Learning Points

An impressive list of common sense checks is the result of this exercise.

- Many checks are possible on a complex piece of equipment such as a car, without any specific expertise.
- Checks do not need to be done every time the car is used, but rather when other work, such as cleaning, takes place.
- Hence, 'Cleaning is Checking'
- Early detection of possible problems – that is Prevention – is made possible.
- Simple checks lead to big improvements in safety, life, and reliability. This is true for almost any item of equipment.
- Just doing it is the key.

The exercise can be extended to include a priority ranking.

## Players

Any number, less than about 20.

## Approximate Time

30 minutes, 1 hour with priority ranking

## Description

Brainstorm a list of simple checks that anyone can carry out on a car.

## Player Instructions

We are going to 'brainstorm' or 'brain dump' a list of checks that anyone can carry out on a car using your 'God-given senses' – or the five senses of sight, touch, smell, hearing, and maybe taste.

A good idea when brainstorming is to begin with a period of silent thought of about 2 minutes. Then, the facilitator will go around the group and ask for ONE suggestion from each. You are allowed to "Pass', but not to criticise any other participant or idea. The facilitator will keep going around until there are no more ideas.

Note: these are not the sophisticated checks that would be done at the time of a major service by a dealer, but the checks that can be done by an ordinary individual with little or no training or expertise.

## Forms, Graphs, Tables, Typical Results

None. You should easily get a list of 25 checks.

## Instructor notes and discussion

Some examples are given in the table. Several more are possible. This table includes some extra categories that can be used to establish priorities. You can use the FMEA categories of Severity, Occurrence, Detection – each weighted 1 nil to 10 high / difficult, to establish the RPN (Risk Priority Number). RPN = R x O x D

## Equipment

None. Paper and pencil exercise.

## References and Further Reading

Bicheno and Holweg, *The Lean Toolbox*, 4th edition, PICSIE, 2009; Pages 70 to 76

## Acknowledgement

Peter Willmot, Willmott consulting, Bath, UK

| Check | Affects | Frequency Needed | Difficulty |
|---|---|---|---|
| Tyre tread depth | Safety, ride | Monthly | Easy |
| Tyre pressure | Safety, wear | Bi-monthly | Moderate |
| Tyre wear evenness | | | |
| Windscreen chips | Safety | Monthly | Easy |
| Oil Level | Wear, reliability | Bi-monthly | Moderate |
| Oil pressure | | | |
| Water | Reliability | Bi-monthly | Moderate |
| Fan belt | | | |
| Screen washer | | | |
| Screen wiper blades | | | |
| Safety belt | | | |
| Headlights clean? | | | |
| Headlights working | | | |
| Brake lights | | | |
| Indicator lights | | | |
| Paintwork | | | |
| Spare wheel | | | |
| Toolbox | | | |
| Loose change store | | | |
| Drinking water | | | |
| Service interval | | | |
| Fuel consumption | | | |
| Cleanliness front | | | |
| Cleanliness rear | | | |
| Flashlight working | | | |
| Rear window demist | | | |
| Air conditioning | | | |
| Front window fog clearance | | | |
| Registration plate | | | |
| Car License Disk | | | |
| MOT validity | | | |
| Key fob battery | | | |
| Battery fluid | | | |
| Electrical connections | | | |
| Wheel alignment | | | |
| Vibration | | | |
| Sound | | | |
| Smell? | | | |

# Spot the Rot

## Key Areas

TPM, maintenance, machines or equipment, asset care

## Manufacturing or Service?

Primarily Manufacturing for TPM purposes but every office has similar opportunities.

## Overview

A simple but powerful exercise to highlight the many, many opportunities that exist for improvement. Go to the Gemba and look.

The inspiration for this exercise comes from Peter Willmott, 'doyen' of TPM professionals in the UK.

## Summary of Learning Points

* A detailed 'go see' exercise reveals many improvement opportunities
* Everyday users often fail to notice opportunities
* Go again, and see more!

## Approximate Time

30 minutes, depending on process

## Description

* This is straightforward
* The group goes to an area or machine and is asked to compile a list of 25 things that are not perfect.
* After 15 minutes the group comes back, are given another sheet, and asked to go back and take another look.

## Players

Almost unlimted. The limitation is the area or machines to be looked at. Each participant must have fair access.

## Player Instructions

As per description

## Forms, Graphs, Tables, Typical Results

Spot the Rot sheet for each participant

**Instructor notes and discussion**

- Almost everyone gets used to less than perfect conditions.
- Fresh and critical eyes see things that often have gone unnoticed.
- A target of 25 is interesting. It may force one to look harder, or to tail-off the observations.
- When prompted with the observations of others, and with a hint sheet, participants invariably find more opportunities.

**Equipment**

Notepads

**References and Further Reading**

- Peter Willmott, *TPM: A Route to World Class Performance*, Butterworth, 2001
- Margaret Heffernan, *Wilful Blindness*, Simon and Schuster, 2012. This entertaining book gives many examples of situations where people look but don't or won't see.

# TV News Production
## *A Game in Lean Design*

**Key Areas**

Lean Design, Lean Project management, Concurrent Design, Cross Functional working, Office Waste, Lead time reduction.

The original game was developed by Kate Mackle of Thinkflow, Lean consultants.

**Manufacturing or Service?**

Manufacturing. However, the example used is of a TV production company.

**Summary of Learning Points**

The exercise can be used as a vehicle to discuss many of the features of the Toyota / Lean Design process. These include: the role of the chief engineer, the role of functional managers, building functional expertise, cross functional understanding, check sheets, leveling or Heijunka in project management, set-based design, 'bookshelving', and the 'Obeya' (Big) room. Features of Goldratt's 'Critical Chain' project management can also be discussed. These include: the critical chain rather than the critical path, slack time at end rather than by activity, but time buffers to prevent non-critical stages holding up critical stages. Lastly, features of 'Lean Construction' project management, whereby check lists are systematically built up and used to ensure activities are not delayed.

- Concurrent working is much more time effective than 'over the wall' production.
- Quality improves, sometimes dramatically.
- Plan Do Check (or Study) Act works well in Design.
- Early decisions lock in many future activities. It is worthwhile spending time up front whilst flexibility is at a maximum.
- The role of the Project Manager: Influence or Authority?
- Being clear on the handoffs from each activity.
- 'Stage Frozen' specifications.
- Communicating with External and Internal customers is important.
- Time over-runs are difficult to control without careful planning.
- Leave slack time at the end, not at each stage.

**Players**

10 minimum, 11 to 16 ideal, up to 20.

**Approximate Time**

2.3 hours, including discussion time. But could extend to 3 hours with longer briefing.

## Description

The team, comprising different departments, have one hour to prepare a 5 minute 'newscast' to be presented from a 'studio'. This is done over two rounds. In the first round, the stages are sequential. Then the team is briefed. They make improvements, decide how they are going to work, and then get 20 minutes to prepare a further 5 minute presentation. It is a lot of fun!

## Player Instructions

The players are divided into sections: Editorial, Research, Writing, Production, News Reading. Two players minimum in each.

The requirements of each section are given on the cards. A collection of current newspapers and magazines must be available. Internet usage, printers, or copiers are not allowed. Laptop computers may be used.

## Forms, Graphs, Tables, Typical Results

Player cards must be given out.

## Instructor notes and discussion

The game requires almost no briefing to start. Just divide up the group and begin. The editors must select the stories and must also specify the requirements as to length and presentation style. Note the general requirements on the story types.

During the first round the Editors may monitor the progress of each stage, but may not talk to the various sections.

Production may be briefed by the Editors in advance, and Production may subsequently set up the 'studio' and talk to the two newsreaders.

Announce that in EXACTLY ONE HOUR the newscast will begin. Be very strict on this. A news broadcast cannot go out late. Warn the Production section 10 minutes before the deadline and then count down minutes from 5, 4, 3, 2, 1 then seconds from 30 to 0. At 0 say 'On Air!' In an actual studio, the last 5 seconds are counted down by showing fingers, and then at zero by pointing with outstretched arm and finger to the newsreader. The instructor should do this. Then, time EXACTLY FIVE MINUTES.

A good sized clock with a second hand or large digital timer showing seconds must be prominently on display for all sections and for the newscast.

The rest of the group should stand around and watch the 'broadcast'. They should be asked beforehand to make notes and prepare comment. They must remain silent during the newscast.

The two newsreaders should begin reading at time 0. Very often the readers will 'run out of steam' before the five minutes. But, they must keep going for the full five minutes. The instructor should, as in a real studio, indicate by making bold circular motions with his or her arms to indicate that the readers must keep going. Then at exactly 5 minutes, the instructor should indicate 'off air' by waving his or her arms boldly in a scissors like motion.

Then ask for comment. Write up the comments on a flipchart. Make points about waste, delay, lack of visibility, taking too much time, and not having enough time, planning, control, communication, understanding what is needed, critical resources.

Then brief the team about some or all of Toyota Design, Critical Chain Project Management, and Lean Construction Project Management. This briefing could be from a few minutes to a full hour.

Features of Toyota Design Practice are given on a slide, but the instructor should read at least one of the references.

Then, say that the Team (or Crew) has 15 minutes to decide on reorganising themselves, and thereafter 20 minutes to prepare for a further 5 minute news broadcast. New news stories must be chosen.

As before, give countdown and signals.

You may allow innovation such as a weatherman with map, and charts and diagrams. The team may give a title to the newscast such as 'Not the 11 o'clock news'.

Finally, discuss the improvements and summarise the lessons. Discuss applicability to own situation.

## Equipment

Pens, Paper pads
A selection of recent newspapers and news magazines such as *The Economist*
Flip chart and coloured pens
A large clear clock with second hand
Five tables
An area for a 'TV studio'
(Possibly) simulated microphones, makeup

Player sheets: Make copies of the Player cards from the CD

## References and Further Reading

- Bicheno and Holweg, *The Lean Toolbox*, 4th edition, PICSIE, 2009; Chapter 15
- Allen C Ward, *Lean Product and Process Development*, LEI, 2007
- Eli Goldratt, *Critical Chain*, Productivity Publishing, 2005
- Ronald Mascitelli, *Mastering Lean Product Development,* Technology Perspectives, 2011 (Highly recommended.)

# Multitasking

**Key Areas:**

Multitasking, productivity, waste, projects

**Manufacturing or Service?**

Both, but very common in office situations

**Overview**

Today, most people multitask. People jump from project to project to emails to meetings, and back to the projects. But at what cost? Delay!

This simulation lets teams of players complete two simple tasks, but in two different ways. The first way is multitasking, the second by completing the tasks in sequence. The deliverables are the same as are the resources. Only the sequence changes. But what about time and errors?

Teams then discuss better ways of working

**Summary of Learning Points**

- Multitasking may appear efficient but productivity suffers.
- Is multitasking more interesting?

**Approximate Time**

30 minutes

**Description**

- The group is divided into teams of five players each.
- Two rounds are played by each team.
- The teams have two A4 sheets per round, plus a red and a blue pen
- The start-to-finish time must be timed by each team for each round

**Players**

Any reasonable number of teams, each with five players – three operators and two 'managers'.

**Player Instructions**

- Each team should have a Multitasking Game Sheet and a Multitasking Instructions
- Each operator does a task in each round
- Follow the instructions

**Forms, Graphs, Tables, Typical Results:** See Game Sheets

## Instructor notes and discussion

- In the first round, ensure that the two managers are focused on their own tasks – red and blue respectively. These managers should shout encouragement to the team when the team is busy with 'their' colour.
- Since the managers would like 'their' projects to be finished as early as possible, they should keep a strict look out for colouring more than 4 at a time of the 'other' colour.
- On a flip chart, draw a table with the results from each team.
- Ask why the difference.
- Discuss the widespread use of multitasking.
- Ask why it is so common.
- What can done in your office?
- In situations such as driving we often multitask: drive, talk, listen to the radio, satnav, think about other things, speak on a phone(?). Numerous studies show that there are real penalties in multitasking in terms of reaction time.
  Research shows that the brain slows down when it has to juggle tasks – but we often don't believe that! It also takes time to re-focus.
- In some cases there are quite dramatic time differences between the rounds, and discussion can get heated.
- Draw a parallel between 'Batch and queue' traditional manufacturing, and one piece flow Lean. In Lean, 'mixed model scheduling' is the way to go in assembly. For instance make A, 2 x B, C, A, 2 x B, C, etc. This is a form of multitasking but don't be confused. These are ongoing operations. It minimises inventory whilst improving service delivery. But each is a complete product. However, in an office or design situation, doing 'mixed model' will hugely extend lead times for completed projects.
- You can discuss the sinister effects of email multitasking. Responding whenever an e mail comes in has at least two problems:
  - ○ It disturbs the flow of thought. Research shows that it takes several minutes to get your mind back to where it was, particularly with thought activities that require high levels of concentration such as design, complex calculations, programming, etc.
  - ○ Priority is given to jobs that are urgent but not important. This is a point made by Stephen Covey. It drives out less urgent but more important work.
- This game can usefully be paired with the Simultaneous Projects Game.

**Equipment**    A4 sheets and pens

## References and Further Reading

- George Gonzalez-Rivas and Linus Larsson, *Far From The Factory*, CRC Press, 2011. See the excellent notes in Chapter 4
- Joseph Hallinan, *Why We Make Mistakes*, Broadway, 2009, Especially Chapter 5
- Stephen Covey, *The Seven Habits of Highly Effective People*, Free Press, 1989

## Acknowledgement

The inspiration for this game comes from my friend Linus Larsson, co-author of a groundbreaking book in Lean service.

# Simultaneous Projects

## Key Areas

Lean design, multi project management, uncertainty

## Manufacturing or Service?

Both. In manufacturing this would be in a design group. In service this could be in consulting engineering, design, advertising, publishing, and many others.

## Overview

The game or simulation takes four simple projects and runs them one, two, and four at a time. Completion times are examined.

Several games are run in parallel, and results compared

## Summary of Learning Points

- Sequential projects lose lots of time waiting for the completion of prior activities.
- Simultaneous projects reduce waits at the cost of extended times for parallel activities.
- Too many parallel projects lead to extended project completion times.
- There is an optimal balance between waste of waiting and completion times.
- Maybe two or three simultaneous projects is best.
- Waiting times are often lost or obscure due to Parkinson's Law: work expands to fill the time available.
- This is a simple case; real project co-ordination increases complexity exponentially!

## Approximate Time

90 minutes, including discussion

## Description

- Refer to the Game Sheet and Player instructions.
- A game board display, or preferably a data projector using PowerPoint is used to display the current period. The instructor may also call out earch period.
- It is possible to play several parallel games. Each game has a minimum of four players.
- Each game has four projects: Red, Blue, Green, Yellow. Each project has four activities 1, 2, 3, and 4. Each activity is undertaken by the same player in each of the projects.
- Each team should have either an A3 size copy of the Game Sheet or should draw out a copy of the game sheet onto a large piece of paper (A3 or A2 size).
- There are coloured discs for each activity. These are moved forward as each project proceeds. There are two discs per project, corresponding to the project colour. At the beginning of each round there will be appropriately coloured discs on Activities 1 and 2 for each project (2 red, 2 blue, 2 green, 2 yellow).
- Each game has three rounds, as shown on the Game Sheet.

- Note the network precedence diagram for a project. Activity 3 cannot begin before activities 1 and 2 are complete. In other words, both discs must be available, indicating completion by both activity 1 and activity 2.
- Activity times are simulated by rolling a dice each period. The activity is complete when a 6 is rolled. This is a realistic time distribution for project activities and follows a Beta distribution. Refer to the slide 'Rolls to obtain a 6' found in the Service Call Centre game.
- The Rounds: In each round, the current period must be on display. Do not move onto the next period until all active players have rolled their dice. In each Round, players 1 and 2 begin in period 1. Players 3 and 4 must wait until discs become available.
    1. The four projects take place sequentially. For instance, Player 1 can move to the next project in the next period as soon as he or she rolls a 6. Likewise player 2. Player 3, of course, must wait for both discs to be available, and then when a 6 is rolled can move onto the next project. Player 4 must wait for player 3 to roll a 6. Player 4 must work sequentially through projects R, B, G, Y.
    2. Players can work on two projects, alternating between them. For instance, player 1 can work on R in period 1 and B in period 2. Where player 1 works during period 3 depends on the first dice roll. If a 6 was rolled he can proceed to project G. If a 1 to 5 was rolled, he works on project R. In period 4, he works on either B or Y, depending on the roll in period 2. And so on. Similarly for player 2. Player 3 must wait for both discs to be available. If two projects are available, he wust alternate between them. If only one is available, he can stay on that project until another becomes available. Player 4 must wait for activity 3 to be completed before starting. Project R must be complete before G can begin. Project B must be complete before project Y can begin.
    3. Players can work on four projects, alternating between them. The alternating rules are similar to Round 2, except alternating between 4 projects. Players 3 and 4 must alternate between all available projects.
- The round proceeds with players following instructions as given on the player instruction sheets.
- For activities 3 and 4 both coloured discs are moved together
- The round is complete when each of the four projects completes activity four.

## Players

- Several parallel games, each with 4 players.
- If there are additional players available let them be recorders.

## Player Instructions

- Follow the instruction sheets
- Player 5 should record the completion times (in elapsed periods) for each project, each round.

**Forms, Graphs, Tables, Typical Results**

- Please refer to game slides

**Instructor notes and discussion**

- The game illustrates, and facilitates discussion about, one of the classic issues in managing project or design offices: How many projects to work on simultaneously?
- The dilemma occurs because of dependent events. This may also occur with 'stage gates'.
- The projects here are unrealistically simple, but are fine to enable the points to be brought out.
- Ron Mascitelli, Lean Design author and expert, discusses such issues in some of his excellent books. Ron says that the optimal number may be three or four. Of course, here we are considering a very simple case and also assuming that each project has similar NPV (net present value). This is unlikely.
- You can extend the game by setting up costs and benefits. This can lead to a good competition between teams, say in a fourth round. Try using these:
  - Project R has a future income of £10,000, decreasing by £50 per period from the start of the game until project completion.
  - Projects B, G, Y have a future income of £5,000, decreasing by £30 per day from the start of the game until project completion.
  - Players cost £100 per player per period.
- Fewer projects allow focus but risk too much attention. Whilst waiting for prior events to complete, work may simply expand to fill the time available (Parkinson's Second Law). But in doing so the project might incur increased, unnecessary complexity. On the other hand, slack time may be used to reduce costs through value engineering.
- This game can be treated as one of a suite of games in the Lean Design or project management area. The other games are
  - The Multitasking Game
  - Lean Design Project Game.
  - The TV Newsroom Game

**Equipment**

- You will need a room with a table for each team.
- Duplicate the game sheet for each team on A3 size paper, or draw out the three projects on a flipchart-sized sheet, for each team.
- Duplicate the project disc sheet and cut out the discs. If possible laminate the sheets before cutting out.
- A dice is required for each team.
- Flipchart for results

**References and Further Reading**

- Ronald Mascitelli, *The Lean Product Development Guide*, Technology Perspectives, 2006

# Lean Design

## Key Areas

Lean design, Project management, customer value, project cost tradeoffs, 3P (Production Preparation Process), Understanding customers

## Manufacturing or Service?

Both: product design or service design. The game is set up to make two engineering-type artifacts.

## Overview

Competitive teams build two artifacts – a tower and a bridge. Costs are associated with the materials and each artifact has a financial measure of performance.

Teams must design and build in a specified time. After the first artifact is built, teams an opportunity to review design and build practice. They are given a checklist for this. Then teams build a more complex artifact. Here there are opportunities to experiment with various Lean design concepts.

## Summary of Learning Points

- Understanding the customer
- The several aspects of what makes a good product or design
- Project planning
- Role of the project manager (Toyota 'Chief Engineer')
- Role of stage gates
- Simultaneous engineering
- Set based design?
- Prototyping
- Building skill in functional areas
- Introduction to 3P (Production Preparation Process)

## Approximate Time

2.5 to 3 hours

## Description

- Teams of 6 (or near) are selected. Try to split skills between the teams. For example, a marketing person, a production person, an accountant, a planner, an engineer in each team. In particular if there are structural engineers or civil engineers in the group, these should be spread between the teams.
- One or two additional people play the role of the supplier. This person must keep a record of all materials supplied to each team.
- No materials are distributed in advance. All materials must be requested from the supplier.
- No carry over of materials from Round one to Round Two

- The instructor plays the part of the customer. The customer answers queries about the designs, likes and dislikes, only if asked. No prompting.
- Teams are each given the Lean Design Project 1, and Materials sheets
- Time 45 minutes, preferably on a data projector showing a count down timer.
- Evaluate and calculate.
- The instructor gives feedback; hands out the 'Questions after Round 1' sheet; and can show the Design Game Evaluation sheet.
- Explain as necessary.
- Teams should NOT be asked for their comments on project management at this stage.
- Teams are then allowed 30 minutes to discuss. Give each team the "Between Round 1 and Round 2' sheet.
- Start Project 2. Give out the design brief ('Lean Design Project 2' sheet).
- There is a 'Stage Gate' after 40 minutes during this round. Each team makes a 2 minute presentation.
- A countdown timer should be used. But note the possibility for extension up to 10 minutes at a cost of £100 per minute.
- Each team presents its bridge.
- Evaluation, testing and comments by the instructor.

## Players

Teams of 5 or 6. Up to 25 players

## Player Instructions

See game sheets. The instructor to lead.

## Forms, Graphs, Tables, Typical Results

Game Sheets

## Instructor notes and discussion

- As the instructor you will have to use your judgment in evaluating and commenting on the various designs. Strike a balance between nice and tough.
- This game can be a lot of fun. Keep it so!
- Criticise teams if they don't speak to you.
- In judging Project 2 there is reward or penalty for height. Note that this should be the loaded height with the small can that you will place or suspend mid span. You may have to use string to suspend the can.

- Column Design. It seems obvious that the best design will be placing cans on top of each other. But wait! Columns fail in one of two ways: material failure and buckling. Material failure is unlikely with both cans and cardboard. But (for any engineers in the group) buckling failure is directly proportional to the moment of inertia, called 'I'. If the moment of inertia of cardboard can be increased by using a made up hollow square, it can be quite competitive because it is cheaper. Packaging tape is expensive. A shorter column made of cardboard may beat a taller column of cans......

- Bridge Design: Several options are given for teams to try out before Round 2 begins. This is to simulate prototyping, the Toyota 'exchange curve' and set based Design concepts. Although exchange curves cannot be developed in the time available (and without engineering knowledge) it should be apparent that different structures are more cost efficient depending on the span. Generally girder, box beam, truss, and suspension. Technical note: A truss could be quite effective. Compression members made out of cardboard would require extra thickness but tension members could be quite thin. A suspension solution need not have a deck like a suspension bridge. String can be used. The sag of the span parabola is a trade off: bigger sag can carry higher load but requires higher towers.

- The group could be introduced to the following concepts.
  All are associated with 'Lean Design:
    - The Double Diamond (Doing the right project, then doing the project right). See the slide.
    - Lean Design is about flow of knowledge. How can knowledge be made to flow in an uninterrupted way?
    - Go see / Go talk / Go to the Gemba
    - Kano Model: What are
        - The basics that just have to be done
        - Performance factors
        - Delighters
    - Set based design. Keeping the options open for some time, not pre-deciding and then discovering that the customer does not like it, or changes his mind.
    - Prototyping and Testing: experimenting just far enough so that a prototype can be shown and (sometimes) market tested before expensive commitment. A variation on set-based design.
    - Exchange Curves. Developing and experimenting so that there is a range of alternatives, with the best one identified for various situations. In this game, this is the idea behind the slide with four hints for spanning a gap. Which is better/cheaper for various gaps or various loads?
    - The Chief Engineer. Toyota uses a chief engineer for new car design. This person has end-to-end responsibility – concept to success or failure in the market. He has responsibility for the car, but no direct authority. He manages by co-ordination, questioning and knowledge.
    - Planning Board. An example is given. This allows responsibilities to be made clear and capacity to be managed. This can be set up in the Obeya (Big Room for project co-ordination).
    - Dimensions of Design evaluation. See the sheet due to James Adams. It is not good enough just to hit the financials.
    - Bart Huthwaite uses the amusing but useful statement that Lean Design is about maximising the 'ilities' (such as performability, affordability, featureability, deliverability, useability, maintainability, durability, etc) whilst minimising the evil 'ings' (such as scheduling, tracking, tooling, inspecting, packaging, documenting, monitoring, etc.).

## Equipment

- Copy the Game Sheets as required.
- There must be sufficient materials available for each team, including
    - Food cans (9 per team). Standard Own Brand baked beans tins (unopened)
    - 5 cm wide plastic packaging tape. Sticky one side. One roll per team.
    - String (allow 3 metres) per team
    - Cardboard A4 size (or approx.). 10 per team. Note any faily stiff cardboard can be used, such as from the backing of note pads, or packing boxes. Note that the longest side must be less than 30 cm.
    - BlueTac (or similar). Two small packs per team
    - Two rulers per team
    - Three pairs of scissors per team
- For Round 2, there should be two tables available per team, spaced 30 cm apart.
- In Round 1, don't forget the 'blow test'. Take a deep breath and blow onto each structure approximately two thirds up from the base. If the column falls down, the team is disqualified.
- The instructor should obtain a small can of some food – like a sardine tin – that can be placed at the middle of the span in Round 2. If the bridge collapses, the team is disqualified. The can should be shown to the teams.

## References and Further Reading

- Ronald Mascitelli, *The Lean Product Development Guide*, Technology Perspectives, 2006
- Bart Huthwaite, *The Lean Design Solution*, Revised edition, Institute for Lean Innovation, 2007
- Rob Westrick and Chris Cooper, *Winning by Design*, Simpler, 2012
- James Adams, *Good Products, Bad Products*, McGraw Hill, 2012

The Lean Games and Simulations Book

# The Happy Pig: An Exercise in Standard Work

## Key Areas

Standard work, standard operations, SOPS, 5S

## Manufacturing or Service?

Manufacturing, some service

## Overview

A quick and fun exercise, bringing out lessons on the importance of standards.

## Summary of Learning Points

- Standards lead to an immediate improvement in quality
- But, no standard is perfect.
- There is always room for improvement.
- An optional opportunity to write a standard operation

## Players

Any. Individual activity.

## Approximate Time

20 minutes. Can extend to 1 hour if players are required to write and test their SOPS.

## Description

Players are simply asked to draw a happy pig. Some drawings from players are shown. Then show how a happy pig can be easily drawn. Note improvements. But, this is not the perfect happy pig. Another round?

## Player Instructions

- Draw a happy pig
- Draw a better happy pig
- Write a standard work instruction on drawing a happy pig.

## Instructor notes and discussion

- Ask participants to draw a picture of a happy pig.
- Ask for volunteers to show their pictures. (Some will embarrassed, so don't push.)
- Note the huge range and variety. Typically from 'sausage dogs' to very good drawings.
- Show the group how to draw a simple, quick happy pig – using concentric circles. See the slide.

- Say that this picture is probably better than most, but is certainly not the best pig that has ever been drawn.
- Note that a standard drawing reduces variation, improves quality, and is often faster.
- A standard builds on past experience, but is never fixed. See the excellent quotation from Henry Ford, 1926.
- A better happy pig is not just a front view. Is the next drawing better?
- Ask the players to re-draw. Compare. Note improvements.
- Show the third drawing. This third drawing IS better.
- Optionally, ask students to make up a standard procedure or (SOP) for drawing a pig. Possibly use the third pig. See annotations.
- Note that a SOP does not just have to be words. There are many applications that just use drawings – with critical points shown.
- This is an opportunity to discuss TWI type work breakdown sheets. These have main points, key steps and reasons for the key steps.

## Equipment

A4 paper sheets
Power point slides

## References and Further Reading

Bicheno and Holweg, *The Lean Toolbox*, 4th edition PICSIE, 2009, pages 82 to 84

For TWI work breakdown, see Patrick Graup and Robert Wrona, *The TWI Workbook,* Productivity Press, 2006

# The Standards Debate

## Key Areas

Standards, SOPS, TWI Job Breakdown, Law of Requisite Variety

## Manufacturing or Service?

Both, but particularly Service

## Overview

There is an ongoing and often heated debate about the applicability of standards, standard operating procedures (SOPS). On the one hand we have statements such as "No improvement without standards', 'Standards are the basis of Quality'. In many representations 'Standards' are the foundation of the 'Temple of Lean'. On the other hand there are statements such as 'Standards cause Failure Demand', and 'Standards don't work in service'. Imai's famous 'Kaizen Flag' diagram shows standards applicable from top to bottom in a 'kaizen' organisation, but with decreasing amount of time spent following standards as the level in the organisation increases.

Many have their own strong views about standards and SOPS. Sometimes it is that 'Standards are good – for the other guys!'

This is an exercise (rather than a game) that opens up discussion about standards in different environments. This is an important discussion for managers. Misapplication of standards can prove disasterous.

The writer is of the opinion that TWI Key Points (or Job Breakdown Sheet) can make a significant contribution. A key point in TWI is an action that 'makes or breaks a job', can lead to injury, or makes the work easier. So some activities would have few if any key points; others need to have many detailed key points.

## Summary of Learning Points

- The applicability of standards differs by situation.
- Misapplication of standards (too much or too little) can cause quality problems or 'failure demand'.
- Key points are a very useful concept. Some situations require few or none, others many key points.
- Ashby's Law of Requisite Variety says that variety is needed to control variety. In an open situation, such as occurs in many service situations, being too prescriptive spells disaster or annoyance. Hence the John Seddon statement that a system needs to have the capability to 'absorb variety'.
- Often flexibility is required, but nevertheless a few key points need to be covered.

## Approximate Time

45 minutes, including discussion

## Description

- Players are each given a sheet to complete.
- Sheets are provided for each player, but only one need be completed.
- Along each 'Job' row, players should place an X at a point on each bar, and make a note about standards.
- A discussion led by the instructor follows.
- Discuss those points where there are significant differences in the location of the X on the bar.

## Players

Any number but a group above about 16 becomes unwieldy.

## Player Instructions

- Use the sheet given out by the instructor.
- Along each 'Job' line, indicate your opinion by placing an X along the line.
- Add your conclusions about standard work in the last column.

## Forms, Graphs, Tables, Typical Results

See Sheet for completion by players

## Instructor notes and discussion

- Note: Standard work and Standardised work are sometimes used interchangeably, sometimes not. At Toyota 'standardised work' means that a process is operating as specified. Thus it is a condition.
- SOPS come in various degrees of complexity. In Pharma companies, for example, they may be extremely detailed, running to 20+ pages.
- Photographs are often incorporated.
- The TWI Job Breakdown sheet has three columns: Main Steps, Key Points, Reasons for the Key Points. This is very good practice for all documentation.
- TWI Job Breakdown does not document every single detailed detailed step – only the 'key points'.
- Opinions differ as to whether operators need to follow the job instruction/SOP or whether they are expected to know the steps by heart. This is a good point for discussion.
- There are other forms: Checklists have been around for a long time for example by airline pilots), but their use is increasing markedly – for example in health care.

## Equipment

None

**References and Further Reading**

- The classic diagram of the 'Kaizen Flag' is in Chapter 1 if Masaaki Imai, *Kaizen*, McGraw Hill, 1986
- Chapter 4 of Masaaki Imai, *Gemba Kaizen*, McGraw Hill, 1997
- See Chapter 1 of Timothy Martin and Jeffrey Bell, *New Horizons in Standardized Work*, CRC Press, 2011. This gives types of manufacturing standardized work
- Joseph Niederstadt, *Standardized Work for Noncyclical Processes*, CRC Press, 2010. Mainly for manufacturing support activities. This gives examples of work documentation. On page 30 is a table showing what types of standard work documentation are required (or not required) for various support activities.
- Joseph Michelli, *The Zappos Experience*, McGraw Hill, 2012. The famous on-line shoe retailer has minimal written standards for callcentre operators.
- Excellent examples of TWI Job Instruction and Job Breakdown Sheet are found in Patrick Graupp and Robert Wrona, *The TWI Workbook*, Productivity Press, 2006
- Atul Gawande, *The Checklist Manifesto*, Profile Books, 2010.
- John Bicheno, *The Service Systems Toolbox*, PICSIE, 2012. Section 4.6 has an extensive discussion.

# The Supplier Partnership Game

## Key Areas

Partnership, supply chain, negotiation, game theory, tit for tat

## Manufacturing or Service?

Manufacturing; some service organisations, like retailers and their suppliers, are strongly involved in this type of situation.

## Overview

This is a classic 'game theory' situation, based on 'the prisoner's dilemma'. Player teams may think they are competing against one another, but in fact both are competing against wider competition. Hence, it is 'supply chains compete', not 'companies compete'.

This game is an ideal 'ice breaker' for a supplier conference or an event where Lean supplier strategy is discussed, or for service operations.

The game is ideally played in two halves. In the first half there is an 'encounter' or 'one-off' transaction between players with no possibility of a repeat transaction. This would be like buying a product or service that is unlikely to be repeated. In the second half the game is played over several rounds enabling the possibility of a 'relationship' to be established.

This game was developed from game theory. Game Theory has been around for several decades, since John von Neumann worked in the area in the 1950's. Many will have seen the movie, *A Beautiful Mind*, about the game theorist John Nash. Nash won a Nobel Prize for the 'Nash Equilibrium' concept.

## Summary of Learning Points

- The game is all about trust and partnership.
- Supply chains compete, not companies: you and your suppliers, not you vs. your suppliers.
- There is a greater payoff with cooperation, but trust is required.
- Steven Covey says, 'Seek win win or walk away'.
- Communication with suppliers is important. This builds trust and understanding.
- There is a temptation to break agreements towards the end of the game (or contract), especially when it is 'one off'.
- Going along with your negotiator, and sticking to agreements.
- Is 'tit for tat' a useful strategy?
- A payoff matrix aids clarification.
- Some forms of co-operation may be illegal – for instance price fixing. These can favour the companies but work against customers.

**Players**

A large number of players can be accommodated. Each game has 3 players, and several simultaneous games can be played. If you have a big group (say 18 or more) then you can double up the number of players to six.

**Approximate Time**

1 hour, longer with discussion on game theory and supplier strategy

**Description**

Players are formed into groups of 3. Each player plays separately 'against' the other two in their game. In other words there are three sub-games per group. Thus
- Player A plays against player B
- Player A plays against player C
- Player B plays against player C

The scenario is one of minimising cost in a transaction (first half), or series of transactions (second half).

The instructor can select an appropriate situation for the players' businesses.  Examples may be

- A shop works with a delivery service. Information about the customer is shared or not. It costs time, effort and money to obtain information. If shared, costs can be reduced; if not costs will be higher for one, lower for the other.
- The 'shop' and 'delivery service' could be replaced by a two building contractors, two travel services (such as train and hotel), an insurance agent and an insurance company, and so on.
- A manufacturer works with a supplier to obtain a component. Information about the end product and the component is shared or not. It costs time, effort and money to obtain information. If shared, costs can be reduced by product modification, less scrap, less warranty claims in future and so on; if not shared, costs will be higher for one, lower for the other.

First Half:

- Explain an appropriate Scenario (as above examples).
- Say that this is a one off situation. Do not give an indication that there will be another round.
- Give each player 20 tokens. A token can be any object – like a piece of paper with a company stamp, or poker chip. Explain that at the end of the first half tokens can be exchanged for something depending on your budget: bottles of beer, pens, candy – but not money. Maybe 5 tokens = 1 beer.
- A token could also just a point score that must be kept by each team. Not quite as much fun.
- It is recommended that a special prize be given to the winner of each group. This should not be a prize that can be easily split between the three players in a group. If this option is

taken, the instructor should announce the prize at the beginning of each half. Do not give money. Perhaps a book or a pen.

- Hand a first half player sheet to each player.
- Explain that no communication is allowed, other than passing a decision slip of paper just once.
- Ask the players to make their decisions. Two decisions per player, one for each of the other players in their group.
- Place the pairs of decisions face down on the table.
- Turn over the decisions and hand over the appropriate number of tokens to the instructor.
- Let each group discuss.
- The instructor leads a discussion.

Second Half

- Explain that, in this second half, there will be a series of decisions, not just one as in the first half. The game will be played over 4 rounds.
- During this second half the players should be separated so that they cannot see the decisions by the third game. For instance, player A will see decisions A and B, and A and C, but should not see the decisions between players B and C.
- The scenario will be the same, except that there will be multiple rounds.
- Give each player 20 tokens.
- Hand players the game sheets for the Second Half.
- Explain that no communication is allowed, other than passing a decision slip of paper just once.
- For each round, ask the players to make their decisions. Two decisions per player, one for each of the other players in their group.
- Place the pairs of decisions face down on the table.
- Turn over the decisions and hand over the appropriate number of tokens to the instructor.
- Repeat for next round, until 4 rounds have been completed.
- Let each group discuss.
- The instructor leads a discussion.

**Player Instructions**

All the game instructions are given on the game sheet. Your instructor will inform you of your group letter and team number.

**Forms**

The player form is given below. Make one copy for each player.

**Instructor notes and discussion**

- Do not tell the players that the game will be in two halves.
- Do not allow talking during the decision making phase.
- For the first half (one-off encounter) the rational or expected choice is that players will choose to not provide information. This is the safe choice for a one-off encounter, because each player stands to lose 0 or 5 tokens.

- There is more risk if a player choses to provide information because the other player may take advantage.
- If the various games have different outcomes, get the best and worst players (most and least tokens remaining) to explain their rationale.
- Sometimes if players know one another they may both independently decide to be 'nice' and to provide information. But can the other person be trusted?
- In real encounters, why take a risk?
- Discuss this risk.
- The value of the prize is a factor. A good prize for the game cannot be split or shared.
- Then announce that a second game will be played over four rounds. Give out the second half sheets. Discussion is allowed before the game starts, but not thereafter.
- In this round, there is a possibility of a trusting relationship. If maintained (both sides elect to share information, and stick to this decision), costs are minimised.
- Of course, trust is required. The partnership lesson is that both parties win so it is not a question of you vs me, but our partnership vs other partnerships.
- Did all players stick with the decision, or were some tempted to break ranks during the fourth round?
- Discussion can be amused, heated, or even angry! Good!
- A 'tit for tat' strategy (as discussed by Robert Axelrod) is often thought best:
  - Be nice at first.
  - In all rounds punish not nice behaviour (eg not providing info) by also not providing info.
  - Be forgiving by offering the other party a chance to resume cooperation.
  - Be clear in the signals you send.
- Of course, in the 'real world' this strategy may not be possible. For instance, there may be a shortage of money, or strong internal incentives or KPIs.
- There is also a danger when relationships (Contracts? Managerial positions?) come to an end or change.
- There is a 'norm of reciprocity' in many relationships: good behaviour attracts good behaviour, and the reverse.

## Equipment

Game sheets photocopied
Small pads for decisions
Card holder for Team names
Data projector

## References and Further Reading

- Macbeth, D.K. and Ferguson, N., *Partnership Sourcing: An Integrated Supply Chain Approach*. Financial Times/Pitman, 1994
- Avinash Dixit and Barry Nalebuff, *The Art of Strategy*, Norton Press, New York, 2008 (This is a wonderful read on game theory but no mention of Lean)
- Ken Binmore, Game Theory: *A Very Short Introduction*, Oxford Univ Press, 2007

155

**The Supplier Game: Player Instructions**

**Objective :** The objective is to minimise cost (in the form of tokens) to your business.

Each player must keep track of his or her own number of 'tokens'. These represent cash. You will frequently have a negative number of tokens, indicating that you have lost cash. You should aim to keep as much 'cash' as possible. The player with the most tokens, or least negative number, wins.

This round will be played once only. Groups of three players are set up.

Each player plays two games, one game each against each of the other two players. These are three separate games. That is: A vs B; A vs C; B vs C.

Players should be separated so that they cannot see the decisions or outcomes of the game between the other two players (Game B vs C, if you are A). No talking or discussion is allowed.

You should record your own score, and the score of the players you play against. Scores are calculated as follows:

| If: | | The Other | Player |
|---|---|---|---|
| | | Don't Provide info | Provide info |
| You | Don't Provide info | You each lose 5 tokens | You don't lose any tokens; the other player loses 10 |
| | Provide info | You lose 10 tokens; the other player loses 0 tokens | You each lose one token |

Write your two decisions (one for each of the other players) on two slips of paper. The decision is either 'Don't provide information' or 'Provide information'. Your two decisions do not necessarily need to be the same.

Place these face down on the table. Make sure it is clear which slip of paper applies to which of the other players.

When all three players have made their decisions there should be 3 pairs of decision slips on the table, face down. When all three pairs are on the table, turn them over one pair at a time.

Note how many tokens you have lost, if any.

**The Supplier Game: Player Instructions: Second Half**

**Objective :** The objective is to minimise cost (in the form of tokens) to your business.

Each player must keep track of his or her own number of 'tokens'. These represent cash. You will frequently have a negative number of tokens, indicating that you have lost cash. You should aim to keep as much 'cash' as possible. The winner is the player with the most tokens, or least negative number, wins.

This game will be played over four rounds.

Groups of three players are set up.

In each round, each player plays two games, one game each against each of the other two players, for a total of 8 games per player. These are separate games. That is: A vs B; A vs C; B vs C.

Players should be separated so that they cannot see the decisions or outcomes of the game between the other two players (Game B vs C, if you are A)

You should record your own cumulative score, and the accumulated scores of the two players you play against. Scores are calculated as follows for each round.

| If: | | The Other | Player |
|---|---|---|---|
| | | Don't Provide info | Provide info |
| **You** | **Don't Provide info** | You each lose 5 tokens | You don't lose any tokens; the other player loses 10 |
| | **Provide info** | You lose 10 tokens; the other player loses 0 tokens | You each lose one token |

**Note : During the last (fourth) round, all scores are doubled. That is, you can lose 0, 2, 10, or 20 tokens.**

Write your two decisions (one for each of the other players) on two slips of paper. The decision is either 'Don't provide information' or 'Provide information'. Your two decisions do not necessarily need to be the same.

For each round, when all three players have made their decisions there should be 3 pairs of decision slips on the table, face down. When all three pairs are on the table, turn them over one pair at a time. Note how many tokens you have lost, if any. The tokens you have lost should be given up to the instructor. Then proceed to the next round until you have completed four rounds.

# Go See, Genchi Genbutsu

**Key Areas**

Go see, Direct Observation, Learning to see, Going to the Gemba

**Manufacturing or Service?**

Both

**Overview**

This short exercise brings home the point that there is no substitute for direct observation.

We often fall into the trap of sitting in an office and assuming that we can remember detail about events, value streams, processes, products, or customer or employee behaviour. This is unwise, because even though we think we know, there are inaccuracies. Memory is fallible. It is part of efficient brain working to remember only detail that is important at the time.

An everyday object – a bicycle – that everyone has seen many times is used. But how well did you observe, and how much can you remember?

**Summary of Learning Points**

- Memory is flawed. You think you can remember, but detail eludes us.
- Go to the Gemba (point of action or interest) and see for yourself!
- Don't sit in your office or conference room and draw up what you think is the process or value stream map!

**Approximate Time**

30 minutes

**Description**

- Ask how many people are familiar with a bicycle. How many are regular bicycle riders?
- Ask everyone to rate their knowledge of a bicycle on a 1 to 10 scales. 1 is no knowledge; 10 is thoroughly familiar.
- Ask each participant to sketch a bicycle, and include as much detail as possible. Give players 10 minutes to do this, but be prepared to extend the time to 15 minutes.
- Distribute the Bicycle Checklist and ask participants to compare the points with their own sketches.
- Ask participants to re-rate their knowledge.
- Ask participants to comment.

**Players**

- Any number

**Player Instructions**

- Use the description above.

**Forms, Graphs, Tables, Typical Results**

- Bicycle Checklist sheet for every participant

**Instructor notes and discussion**

- Lean puts emphasis on 'Go to Gemba', 'Go See', 'Direct Observation', Genchi Genbutsu, with good reason.
- Inaccurate, imprecise information is probably one of the biggest reasons for Lean failure – and for problems in both society and personal life.
- The brain is a very efficient organ. One reason is that it stores only the essential information necessary for day to day working. If everything 'seen' had to be stored, few brains could cope. But then, sometimes, it is necessary to recall exact information. The brain is then fallible.
- Few, if any, have 'photographic memory.'
- In TWI job relations there is the expression 'get the facts before making a judgment'. So, don't rely on second hand information. Go and see for yourself. Remember also, that everyone has 'filters' – they see what they want to see.

**Equipment**

None

**References and Further Reading**

- Going to the Gemba, and Direct Observation is mentioned in many books on Lean. See, for example, Liker and Convis, *The Toyota Way to Lean Leadership*, McGraw Hill, 2012
- The IDEO Design company is famous for their direct observation approach. See their web site.
- The inspiration of this exercise comes from an experiment by Rebecca Lawson as described in the fun but important book by Christopher Chabris and Daniel Simons, *The Invisible Gorilla*, Harper, 2011. See Chapter 4,
- Joseph Hallinan, *Why We Make Mistakes*, Broadway, 2009, Especially Chapter 1. You will be horrified by some of the examples in this book!

# Targets and Measures

## Key Areas

Inappropriate targets, Motivation, Understanding variation, SPC, control limits, Regression to the mean, Deming

## Manufacturing or Service?

Both

## Overview

This game or simulation illustrates the fallacies of unrealistic target setting, and the tendency of a good proportion of managers to reward inappropriately 'good' performance, and punish 'poor' performance.

The game also gives an introduction to statistical process control (SPC).

Dr. W Edwards Deming developed and played the famous Red Bead game in the 1970's. This game is a simplified version requiring only a pair of dice, but hopefully bringing home most of Deming's message.

An 'obvious' game, but a profound message.

## Summary of Learning Points

- Setting targets (or 'stretch targets') without knowledge of natural variation is not only wrong but can also be de-motivating
- Natural variation of the process should be studied to establish the control limits.
- Deming spoke about the 95:5 rule whereby most problems fundamentally lie with the system rather than the 'workers'. Targets are useless unless the process can be changed. And, generally, managers are the only ones who can sanction a process change.
- Regression to the mean: this is the tendency of apparently good performance to be followed by apparently poor performance.
- What managers 'learn' (wrongly) is not to praise good performance but to punish poor performance.

## Approximate Time

1 hour

## Description

- Select a team of 8 volunteers (6 workers, one inspector, one scribe) or for a smaller group of 6 volunteers (4 workers, one inspector, one scribe).
- Workers do not receive any game sheets.
- The instructor plays the role of the manager.
- Use a pair of dice, the bigger the dice better.

- The instructor explains that the game is a simulation of …… here, select an appropriate situation such as 'sales force', 'field repairs', 'manufacturing defects', 'student failures', 'insurance investigations', 'innovations'.
- The simulation takes place over 4 rounds (select days, weeks, months).
- The manager explains that he has set a target of 9 for each worker, each period.
- Each round, each worker rolls the pair of dice, and adds the two numbers. This is verified by the 'inspector' and written up on a chart by the scribe. The format of the chart is shown on a slide.
- After each round, the scribe adds up the total score and calculates the average.
- Then the manager comments on the round. He praises workers getting top scores and says he is disappointed and angry with those getting the low scores. If any worker gets a score of 9 or more, the manager comments that it proves the target is possible. Low scoring players are told to try harder.
- The next round is then run. The average so far is calculated. Same sequence of comments. Except now the manager, in looking at the scores:
  - If performance of a worker improves he should say that it is due to the severe words expressed in the previous round
  - If performance of a worker declines he should say that 'workers take advantage of praise'
  - If an average so far is near of above 9 he should say that consistent high performance is possible, and encourage the others to emulate it.
- This charade goes on for four rounds after which the totals and averages over 4 rounds are calculated.
- A discussion on managers, measures, natural variation and control limits follows.
- The Control Limits sheet may now be distributed.

**Players**

6 or 8

**Player Instructions**

All instructions are given verbally by the instructor.

**Forms, Graphs, Tables, Typical Results**

See Appendix for

- Targets and Measures Game: control limits
- Motivation and Measures Game: sample and blank sheets

**Instructor notes and discussion**

- Of course, the game is obviously a charade. The results are 'obvious'.
- Nevertheless, many managers persist with such behaviour – sometimes in ignorance.
- While playing the game, the instructor should not allow discussion although cheers of delight and cries of disappointment enliven the game.
- The results of two dice rolls have a predictable distribution, from which the probability of getting a 9 or more can be calculated. It is 6/21 for two rolls.

- For a set of 4 rolls by a number of players it is better to calculate the upper and lower control limits, UCL and LCL. For this we use calculations for a 'p chart' (or percentage) chart, based on a sample we have just performed.
- The tables show UCL and LCL for various total scores, for 4 and 6 workers.
- As long as the worker averages fall within these limits, the worker's performance is common cause variation. No action should be taken; it is just natural variation.
- You will notice the total scores that would be required for an average of 9 to be 'common cause'. Such scores would be highly unusual.
- 'Stretch' targets are, of course, useless – as would be an incentive bonus. The only action that can improve performance is to change the dice to others with higher numbers. Workers cannot do this without management approval, which is improving the process, not managing by incentives or punishment.
- You might discuss how many managers, bankers, stockbrokers, salesmen, etc get a bonus which is actually is achieved through natural variation?
- Of course, truly superior performance (achieved at a point above the UCL) could well be rewarded.
- Regression to the mean is the tendency for a low score to follow a high score. This is simple statistics. For example if a 10 is obtained from two dice rolls, the probability of getting a total of 10 or more for a second pair of rolls is 4/21 (19%) whereas the probability of getting a total of less than 10 is 81%.
- The classic Deming Red Bead game uses a bucket containing red and white beads. A paddle with 50 slots is used by each worker to draw beads. Reds are defectives. These are recorded. After 3 rounds, the 'worst' workers are 'fired', and the last round uses only the remaining 'good' workers. Performance is often worse or the same.

## Equipment

Two Dice
Flip chart

## References and Further Reading

- W Edwards Deming, *Out of the Crisis*, MIT Press, 2000
- Joyce Orsini, *The Essential Deming*, McGraw Hill, 2013. Especially Chapter 7 'Management is prediction: Statistical thinking is required.'
- Donald Wheeler, *Understanding Variation*, Second edition, SPC Press, 2000.
- The Red Bead Experiment can be purchased from The Deming Society.

# Diversity and 'Wisdom of Crowds'

**Key Areas:**

Participation, Respect, Knowledge

**Manufacturing or Service?**

Both

**Overview**

A simple experiment to show the power of group decision-making.

The bestselling book 'The Wisdom of Crowds, Why the Many are Smarter than the Few' by James Surowiecki gives numerous stories about where a 'crowd', collectively, 'knows' better than an 'expert'.

Of course, this is not always the case. A crowd should not be expected to make a better decision than, for example, a scientist, an engineer, or a doctor, where specific technical knowledge is required. But In Lean there are many situations of uncertainty where judgment is required and where the decision relies on experience rather than technical expertise. Participants must have some familiarity with the situation or technology. Here group or 'crowd' decision-making may be much more effective. Moreover, it shows respect for the participants – something that every Lean transformation should cultivate and encourage.

The group is asked to estimate the number of objects in a jar. Discussion is not allowed. The estimates are written on scraps of paper. The estimates are then written up on a flip chart, and the average calculated.

Two members of the group are asked to count the objects. The answer is compared with the calculated average. The calculated average is often (but not invariably) quite close to the actual object count.

The calculated average is almost invariably better than the majority of individual estimates.

**Summary of Learning Points**

- A group often makes a better decision than most if not all individuals.
- This is the power of diversity.
- Excluding the outliers often will make the group decision worse!
- Conditions where group decision making gives good and not so good results.

**Approximate Time**

30 minutes

**Players**

Minimum of about 8 for fair results. Obviously a larger group takes longer for the jar to be examined and for results to be calculated, so perhaps limit the number of players to about 25.

**Player Instructions**

- Pass around the jar containing the objects.
- Players may look but not touch the objects.
- No discussion is allowed.
- Players write their own estimate of the number of objects on individual pieces of paper. Fold the papers. Do not let other players see your estimate.
- Hand the folded slips to the instructor.
- The estimates are then written up on a flipchart by a group member.

**Typical Results**

Actual estimates for a group of 15 are as follows:

350, 400, 480, 140, 170, 280, 200, 385, 180, 300, 150, 255, 250, 350, 188
or ascending
140, 150, 170, 180, 180, 200, 250, 255, 280, 300, 350, 350, 385, 400, 480

The calculated average of this set is 271.3.

The actual number of objects in the jar was 278.

Notice
- that only one member of the group estimated the number more accurately than the group average. Sometimes the group does better than any member.
- that one member's estimate was within 2 of the actual. The calculated average was within 7 of the actual.
- that the next best player's estimate was 12 from the actual.
- that dropping the highest and lowest estimates will make the average worse. It becomes 265.3. So even 'bad' estimates have value. This is the power of diversity.

**Instructor notes and discussion**

- Do not let the group talk during the estimation phase.
- Ask two group members to write up the estimates on a flip chart, one to write, the other to check.
- After writing up the results, ask the flip chart writers to re-write the estimates in ascending order.
- When writing up begins, ask two others to calculate the average.
- Comment on the answers. See typical results above. Of course, sometimes even better results than the typical will be obtained, but sometimes worse.
- This is an illustration that a group will often make a better decision than most if not all individuals.

- The implications are that everyone should contribute to many decisions. The 'boss' or 'expert' does not necessarily have the best answer.
- This should be salutary for bosses, and encouraging for juniors. It is important in Lean!
- Many similar examples are given in the excellent book 'The Wisdom of Crowds'. Try to read this beforehand. It is both entertaining and instructive.
- There are many, many situations where a representative cross section will do very well – better than 'experts': inflation, stock prices, politics, sales, advertising, are examples. But, according to Suroweicki, the preconditions for a crowd being smarter than an individual are:
  - Diversity within the crowd – several different viewpoints, disciplines, areas should be represented
  - Decentralisation – not, for instance, using only people from one office or area
  - Independence – people must be free to express their own opinions, not influenced by the boss or by fear.
- Calculation can help. For instance if you know the volume of a lego brick, and estimate the 'void ratio': bricks volume to total volume. Such calculations can also be used in Lean decision making.
- An extension: for a larger group, split the group into two sub-groups and perform the experiment. Compare the results as above. If both groups reach a similar conclusion confidence grows. A small risk!
- The same applies for say, market research, where two or more surveys should be compared.
- Group predictions, estimates, forecasts can be improved by giving feedback. A sort of PDSA.

## Equipment

- Two calculators are required.
- You will need to find a suitable, preferably glass, jar.
- The jar should be filled with similar objects.
- The writer uses:
  - A glass jar of 1 litre, with a lid. A vase can be good.
  - The jar preferably has curved sides.
  - The objects are 4 stud lego bricks.
  - Other possible objects: marbles are also OK, nuts (as in nuts and bolts), small wrapped sweets (this has the bonus of giving players a reward at the end of the game!), screws – select as long as possible, and reduce the container size. Avoid using objects like poker chips that can accumulate in columns.

## References and Further Reading

- James Surowiecki, *The Wisdom of Crowds*, Abacus, 2004. Fascinating and midblowing.
- Scott Page, *Diversity and Complexity*, Princeton, 2010. More sophisticated examples, for example, predicting CD rental borrowing. The diverse group often brings multiple 'system' perspectives to complex decisions resulting in remarkable performance. Much fascinating material on group composition, and improving estimates.
- A fascinating description of the wisdom of animal and insect swarms is given in Peter Miller, *Smart Swarm*, Collins, 2010 (People show swarm behaviour also – see the final chapter!)

# APPENDIX

## These slides are available in pdf format from
bichenojohn@me.com
## £20 plus VAT with payment via PayPal

| | |
|---|---|
| **A3** | 167 |
| **Airplane Games** | 176 |
| **Square Games** | 203 |
| **Dice Games** | 223 |
| **Kanban Games** | 234 |
| **Cut & Fill Games** | 245 |
| **5S** | 259 |
| **Manufacturing Cell** | 268 |
| **Paperwork Office** | 272 |
| **Service Call Centre Operations** | 283 |
| **Spot the Rot** | 294 |
| **TV Game** | 295 |
| **Multitasking** | 309 |
| **Lean Design** | 320 |
| **Happy Pig** | 330 |
| **Standards Debate** | 336 |
| **Go See** | 337 |
| **Targets & Measures** | 339 |

# The A3 Improvement Game

Information Sheets (To be cut into strips before the game, and distributed individually only at the specific request of players. Distribute one strip at a time to answer a question.)

## Profits

| Month | Forecast | Actual |
|---|---|---|
| April | £3500 | £4000 |
| May | £5500 | £6100 |
| June | £8500 | £6800 |
| July | £ 9400 | £7000 |

## Types of Problems (Broad) (Customer Survey: Number of Complaints)

| Complaint type | Number |
|---|---|
| Environment | 18 |
| Price | 35 |
| Standard of Service | 36 |
| Product Quality | 97 |
| Staff | 26 |
| Other | 19 |

## Problem types (For Product Quality)

| Quality Problem Type | Frequency |
|---|---|
| Too weak | 9 |
| Cold | 35 |
| Not full | 4 |
| Other | 7 |

## Shift Patterns

| Shift | |
|---|---|
| Morning | 6:30 am to 13:00 |
| Afternoon Evening | 13:00 to 19:00 |

**Staff Types**

| Shift | |
|---|---|
| Morning | Mainly mature ladies |
| Afternoon | Mainly students |

**Problem types (For Standard of Service)**

| Service Problem Type | Frequency |
|---|---|
| Slow | 15 |
| Courtesy | 4 |
| Wrong Change | 3 |
| Other | 3 |

**Complaints (by customer age)**

| Age Group | Frequency |
|---|---|
| Young | 27 |
| Middle | 28 |
| Older | 14 |

**Problems by Time of Day**

| Time of Day | Problem Frequency |
|---|---|
| Breakfast | 8 |
| Morning | 2 |
| Lunch time | 16 |
| Afternoon / Evening | 8 |

**Complaints (by coffee types)**

| Coffee Type | Frequency |
|---|---|
| Filter | 3 |
| Espresso | 4 |
| Double Espresso | 0 |
| Cappuccino | 34 |

**Problems with cappuccino**

| Cappuccino problem | Frequency |
|---|---|
| Too much milk | 2 |
| Cold | 27 |
| Froth | 8 |
| Chocolate added | 4 |

**Process steps (General coffee)**

- Take order
- Place cup in machine
- Check for sufficient beans and water
- Switch on machine
- Enter code for coffee type
- (Water boils / machine makes drink)
- Coffee pours into cup
- Place cup on saucer and serve
- Explain about location of milk and sugar

**Process steps (Cappuccino)**

- Take order
- Place cup in machine
- Switch on machine
- Enter code for espresso
- (Water boils / machine makes drink)
- Espresso essence is added to the cup
- Steam the milk
- Add steamed milk to espresso
- Froth milk if necessary
- Add a 20mm layer of milk foam
- Sprinkle chocolate
- Place cup on saucer and serve
- Explain about location of sugar

**A3 Coffee Shop Exercise: Layout**

The coffee shop has a sit-down area facing onto the street. This has 8 tables (each 1m x 1m), and 20 chairs. The counter is towards the back of the shop. The payment till is on the right hand side of the counter as customers walk in. Customers go to the left of the counter, and pay at the right hand side of the counter where there is a cash till. There is a small selection of cakes on display, and two coffee machines are located on a table along the back wall. Materials (cups, saucers, coffee, tea, trays, chocolate, etc) are stored on shelves along the back wall. Sugar, spoons and small milk cartons are kept at the right hand side of the counter after the cash till. Other milk is kept in the kitchen fridge and brought out in 1 litre jugs.

There is a very small kitchen accessed through a door to the left of the counter. In the kitchen is a sink, dishwasher, fridge, and various storage cupboards. There is also a small hot pan used for heating milk.

**Staff Training**

There is no formal training program. New employees learn the methods from existing staff. 'Just watch me and see what I do'. This takes place in less busy periods.

**Work Patterns.**

One employee stands at the cash till.

All other employees move along the counter (from customer's left to right). They take orders on the left side, make the drinks, and place the dinks on a tray. Employees write a slip with the order and hand it to the cashier.

Customers pick up the tray after paying, and go to empty seats.

An employee clears used cups and trays, takes them to the kitchen, washes them, and returns them to the shelves.

**Students and Ladies**

The 'students' are mainly sixth form but some are undergraduates from a nearby university. The mature ladies are women with kids. They take kids to school in the morning, then work in the shop, then pick up their kids from school in afternoon.

'Mainly' means that a few ladies occasionally work in the afternoon, and one or two students work in the morning.

**Staff stability**

The ladies are stable employees. Students are not. This is mainly because they are in sixth form for less than a year, and go on holiday. Likewise the undergraduates.

**Detailed look at Cappuccino making in the morning**

Milk is sometimes not steamed, but kept in the fridge and then added to the espresso portions. Milk is poured straight from the jug.

**A3 Coffee Shop Exercise:  Types of Coffee**

| Coffee Type | Strength | Useage | Frequency of use |
|---|---|---|---|
| CC2 | 2 | Other coffee types | 2 |
| CC3 | 3 | Standard for filter | 8 |
| CC4 | 6 | Standard for cappuccino | 6 |
| CC5 | 5 | Other coffee types | 1 |
| SB3 | 3 | Standard for filter | 5 |
| SB4 | 6 | Standard for cappuccino | 5 |

**Coffee Suppliers**

| Supplier | Notes |
|---|---|
| CC | Reliable |
| SB | Erratic – often late |

**Customer Coffee Preference**

| Coffee Supplier (Brand) | Customer Preference |
|---|---|
| CC | 1 |
| SB | 2 |

**A3 Coffee Shop Exercise Forms:**

**Coffee Shop Briefing Document**

You are have been asked to assist the manager of a small coffee shop to improve performance.

The coffee shop is not part of a chain, but serves different types of coffee all day long to local shoppers and town visitors. The shop serves coffees and cakes.

The coffee shop has been open for just over two years. Custom has grown steadily. It has become established as a meeting place for adults.

The shop is situated in a small rural market town. There are a few 'fast food' outlets in the town (some serving cheap coffee as a take away), a small hotel, and a few restaurants. At present there no other coffee shops in town including no 'Starbucks' or 'Costa'.

An average of around 250 customers come into the shop each day, but this is unevenly spread. More customers come in on Saturdays. The peak times are breakfast time (highest) then lunchtime. During morning peak most coffees are take-away, but at other times customers sit down and chat over coffee.

Staff turnover is moderate. Often high school girls and boys are employed. They often work at the shop for a few months before leaving.

All new employees are given basic training of two hours, plus are shadowed by a more experienced employee for the first week.

Prices are considered reasonable by customers, and in line with coffee prices in nearby towns.

Recently there have complaints about the quality of the coffee. The profit forecast has been missed.

**A3 Coffee Shop: Mentor Guidance Sheet**

The mentor should not join in with the actual analysis, but should use this sheet to guide the process.

Note and acknowledgement: Many of these questions are from Sobek and Smalley, *Understanding A3 Thinking*, CRC Press, 2008

- Is there a clear theme?

- Are the objectives of the study clear?

- Is current condition clear?

- Has Pareto analysis (80/20) been used?

- How could it be made clearer?

- What is the actual problem?

- Are the facts clear?

- Is there evidence of root case (5 whys)?

- Have the 'six honest serving men' been used? (What, Why, When, Where, How, Who)

- Has the root cause been demonstrated?

- Have all the relevant factors been considered? (Human, Machine, Method, Material, Measure, Mother Nature)

- Has the cause been demonstrated?

- There may be several issues. Which is the big one?

- Are there clear countermeasures?

- Will the actions prevent recurrence?

- How will the results be verified?

# A3 Improvement : Standard Format

| Plant name | Problem type | 4 weeks trend | | | Date | Final Sign Off | | | |
|---|---|---|---|---|---|---|---|---|---|
| | | DT | SPEED | QUAL | | FGM | Author | Manager | CI FAC |
| | | | | | | | | | *John Bicheno* |

**STEP 1) CLARIFICATION OF THE PROBLEM**

**A) BACKGROUND** (problem background explanation, business goals etc)

**B) PROBLEM** (Visualise gap to target)

ULTIMATE GOAL

IDEAL CONDITION

CURRENT CONDITION

PROBLEM STATEMENT

**C) CONTAINMENT** (Describe actions put in place)

**STEP 2) BREAKDOWN THE PROBLEM (GRASPING THE CURRENT SITUATION** (QC tools / process flow / fishbone / go to gemba))

What

When

Where

Who

How

Process study (See by yourself)

SUMMARY OF PROBLEMS TO TACKLE

**STEP 3) TARGET SETTING** (SMART target, graphically shows effect on gap to target from step 1B)

**\*\*Stage 1 Review**     Manager     Author

**STEP 4) CAUSE ANALYSIS**

WHY? Therefore   WHY? Therefore   WHY? Therefore   WHY? Therefore

MAIN

MATERIAL

MACHINE

METHOD

ENVIRONMENT

**STEPS 5 AND 6) DEVELOP COUNTERMEASURES AND EXECUTE**

| Ref | Root Cause | Countermeasure | Evaluation O ▲ X | | | | | | Schedule | | |
|---|---|---|---|---|---|---|---|---|---|---|---|
| | | | Quality | L/T | Cost | Risk | Overall | Priority | Who | Schedule | Status |
| | | | | | | | | | | | |

O= Good   X = None
▲ = Fair   • = Effect = contribution to target

**\*\*Stage 2 Review and approval of countermeasures**     Manager     Author

**STEP 7) MONITOR RESULTS AND PROCESSES**     (Test
Understanding, root cause)

**STEP 8) STANDARDISE AND SHARE SUCCESSFUL PROCESSES**

**\*\*Stage three Review**     Manager     Author

# A3 Improvement : Coffee Shop : Sample

| Plant name | Problem type | 4 weeks trend | | | | Date | Final Sign Off | | | |
|---|---|---|---|---|---|---|---|---|---|---|
| COFFEE SHOP | QUALITY | DT | SPEED | QUAL | / / | | PGM | Author | Manager | CI FAC |

**STEP 1) CLARIFICATION OF THE PROBLEM**
A) BACKGROUND (problem background explanation, business goals etc)

COMPLAINTS

FORECAST ACTUAL. MISSED FORECAST

£2.5K GAP

B) PROBLEM (Visualise gap to target)

ULTIMATE GOAL £10K PROFIT SUSTAINED
IDEAL CONDITION
CURRENT CONDITION £2.5 SHORT
PROBLEM STATEMENT SUSTA(INE) HIGH PROFIT

C) CONTAINMENT (Describe actions put in place)

SPECIAL OFFERS WHILST CLOSING THE GAP

**STEP 2) BREAKDOWN THE PROBLEM** (GRASPING THE CURRENT SITUATION (QC tools / process flow / fishbone / go to genba))

PROBLEM TYPES

PRODUCT QUALITY — SERV. — PRICE — STAFF — FILTER... — COLD — WEAK...

CAPPA

QUALITY CAPPACINO
When MORNING
Where
Who STUDENTS?
How

Process study (Sort by yourself)

TAKE ORDER → PLACE IN MACH. → ENTER CODE → ADD ESPR. → STEAM THE MILK → ADD STEAMED MILK → FROTH → SERVE

HOT COLD COLD

PROBLEM WITH CAPPACINO WHEN MILK IS ADDED

**SUMMARY OF PROBLEMS TO TACKLE**

PROBLEM WITH CAPPACINO IN THE MORNING.

**STEP 3) TARGET SETTING** (SMART target, graphically shows effect on gap to target from step 1E)

DETERMINE REASONS FOR COLD CAPPACINO IN THE MORNING. (LITTLE OR NO PROBLEM WITH OTHER COFFEES)

Manager        Author        **Stage 1 Review**

**STEP 4) CAUSE ANALYSIS**

Why? Therefore Why? Therefore Why? Therefore Why? Therefore Why? Therefore Why? Therefore Why?

COLD CAPPA IN MORNING

STAFF → STU-DENT → NO TRAINING → HIGH TURNOVER → NOT SPECIFICALLY CATERED FOR.

NO STANDARD MEASURE

MEN
MATERIAL  MILK → TOO MUCH?
MACHINE
METHOD  MILK NOT STEAMED → UNDER PRESSURE AT PEAK
ENVIRONMENT

**STEPS 5 AND 6) DEVELOP COUNTERMEASURES AND EXECUTE**

| Root Cause | Countermeasure | Evaluation O△X | | | | | | Who | Schedule |
|---|---|---|---|---|---|---|---|---|---|
| | | Quality | LT | Cost | Risk | Overall | Priority | | |
| NO TRAINING NO JOB STANDARD | DEVELOP J.I. TYPE A TRAINER | O | | O | X | △ | 1 | | |
| NO MILK MEASURE | PROVIDE ONE! | O | | O | O | O | 2 | | |
| | | | | | | O | O | 1 | | |

O= Good  △ = Fair  X = None  * Effect = contribution to target  **Stage 2 Review and approval of countermeasures**

Manager (Test)

**STEP 7) MONITOR RESULTS AND PROCESSES**
Understanding, root cause)

CHECK SAMPLES OF CAPPACINO IN MORNINGS DAILY THEN WEEKLY

Manager        **Stage 3 Review**

**STEP 8) STANDARDISE AND SHARE SUCCESSFUL PROCESSES**

GIVE INSTRUCTION ON BASIC POINTS OF JOB
INSTRUCTION
~ STEPS
~ KEY POINTS
~ REASONS FOR KEY.

Author

# Airplane Games

# Operation 1: Fold 1

A4 sheet

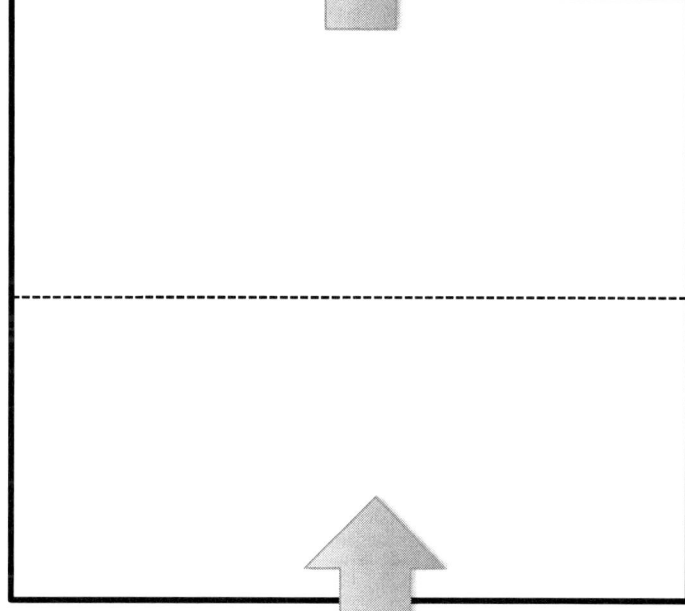

Fold in half

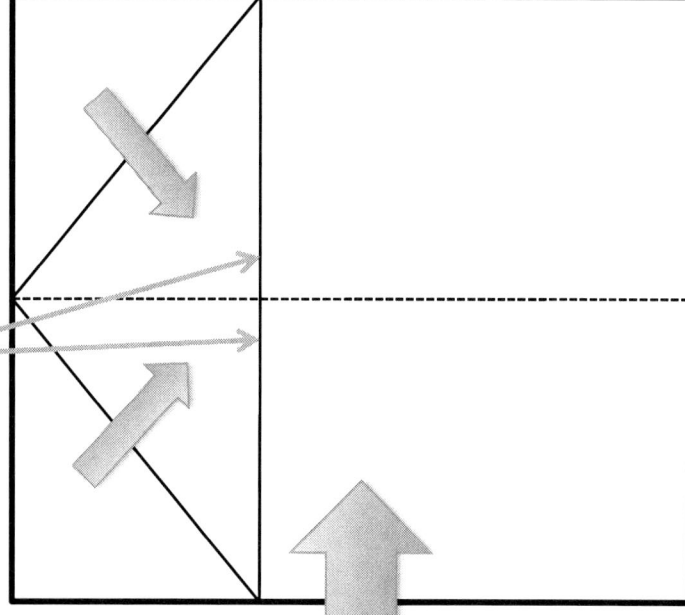

Note:
Align edges

Fold down corners

# Operation 2:  Fold 2

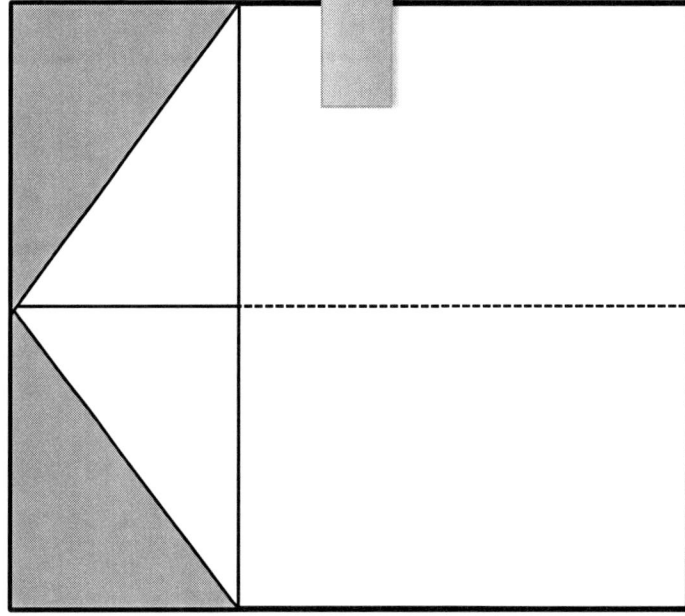

Note:
Align corners

Make second Fold

# Operation 3:
# Folds 3

Pen and ruler required

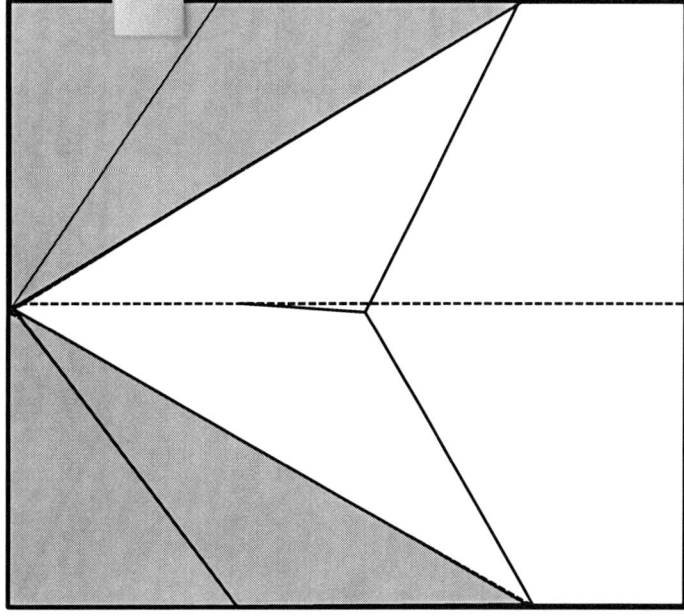

Turn sheet
Thru 90 degrees.
Fold in half along
Centre fold

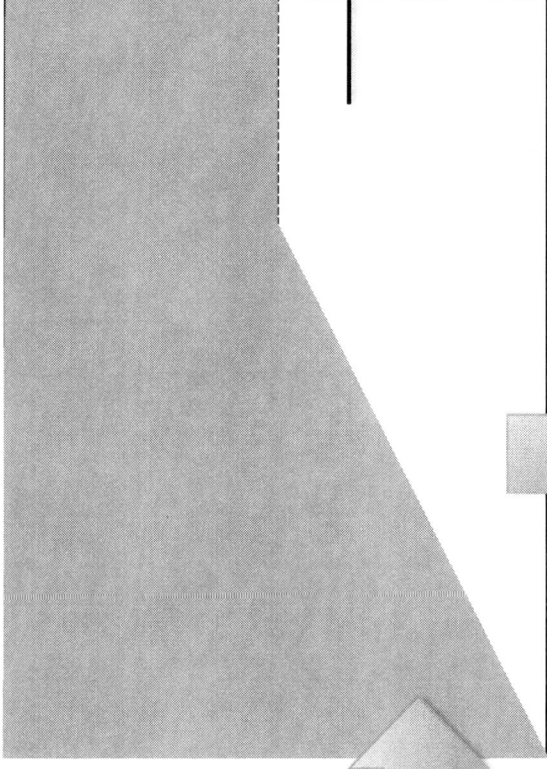

Mark off 1.5 cm
On each side

Fold the wings
over the body
on both sides.
Note 1 cm.

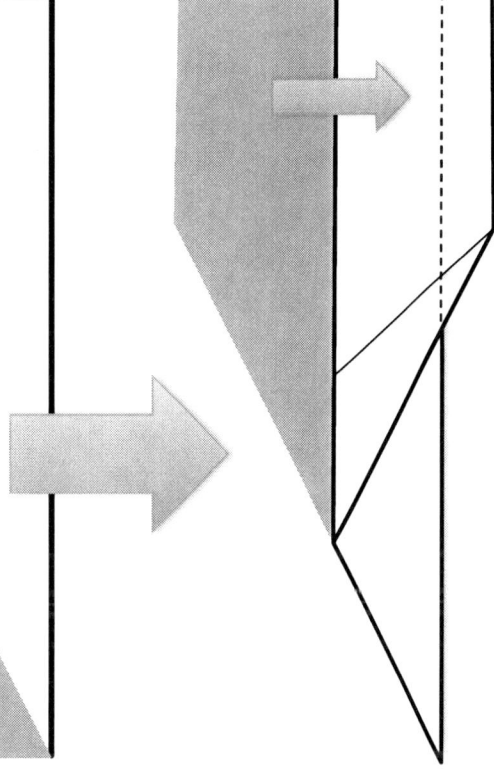

1.5 cm.

Top (sheet) View

Side Views

# Operation 4: Folds 4

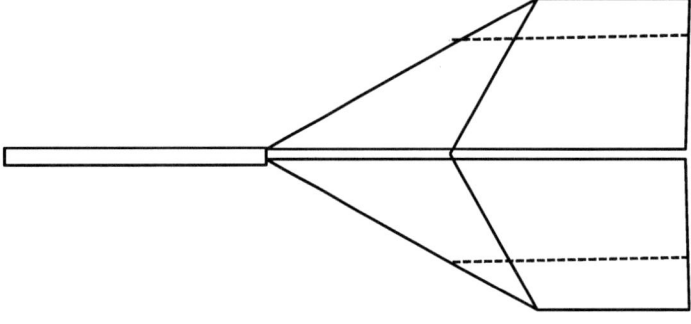

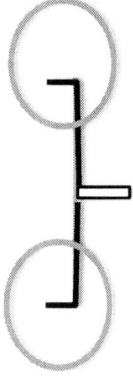

End
View

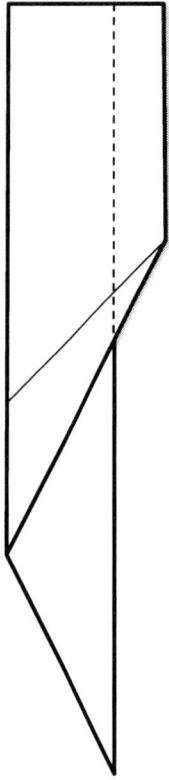

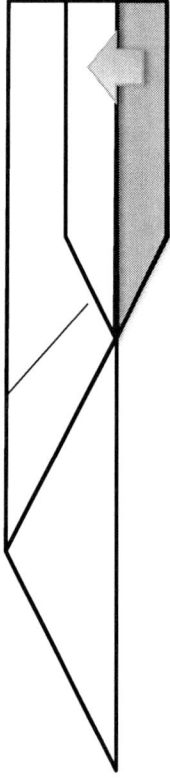

Top View
(when opened)

Fold up both
Wing tips.

Side View

# Operation 5

Red and Blue Pen required : (Only one colour for Basic JIT Game)

Side Views

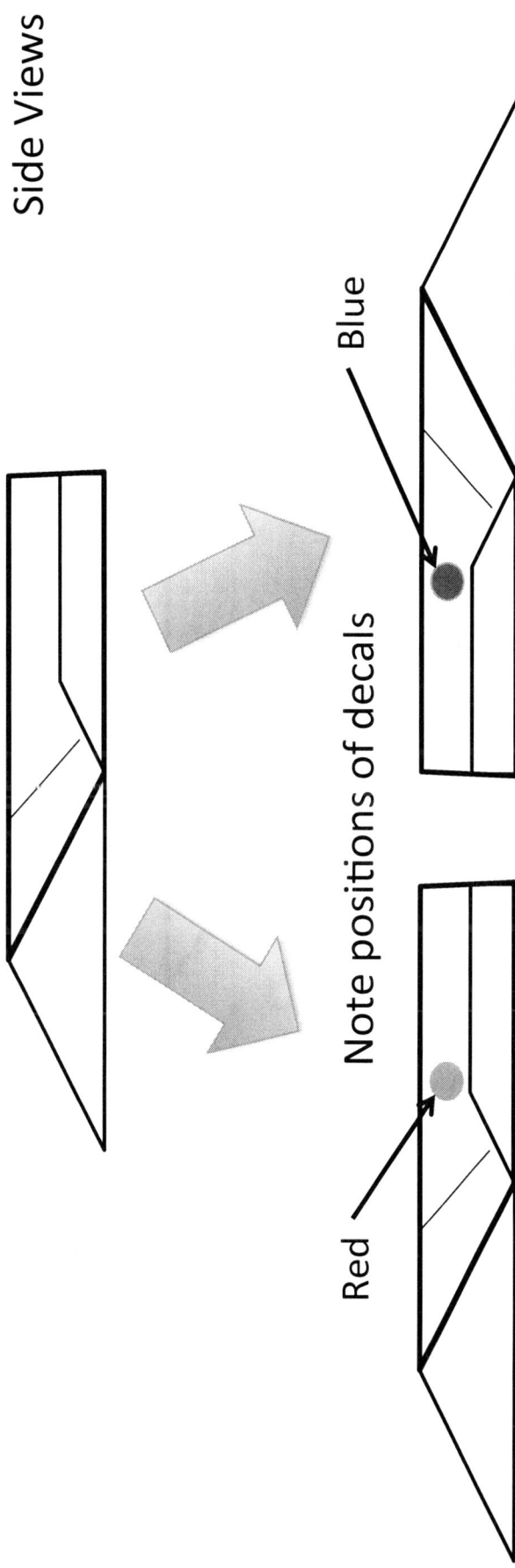

Note positions of decals

Blue

Red

With red pen,
add 1 cm diameter
circular decal to
Left wing of plane

... and with Blue pen,
add 1 cm diameter
circular decal to
Right wing of plane

# Operation 6

Red and Blue Pen required: (Only one colour for Basic JIT Game)

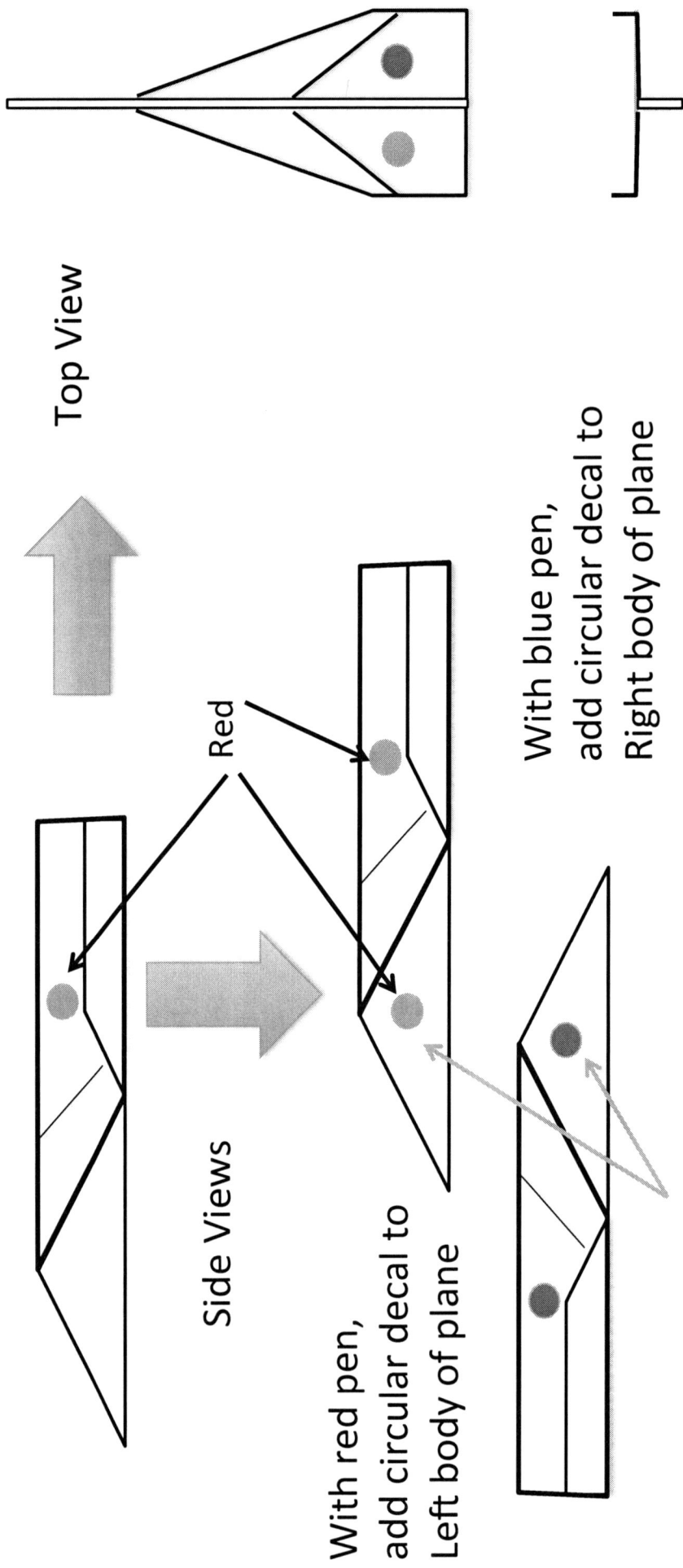

Top View

Side Views

Red

With red pen,
add circular decal to
Left body of plane

With blue pen,
add circular decal to
Right body of plane

Note positions of decals

# Operation 7

Final
Top View

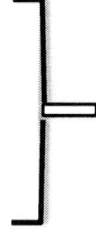

End View

Red Decals

Blue Decals

...Open up wings, straighten all edges. Then add three staples along the fuselage.

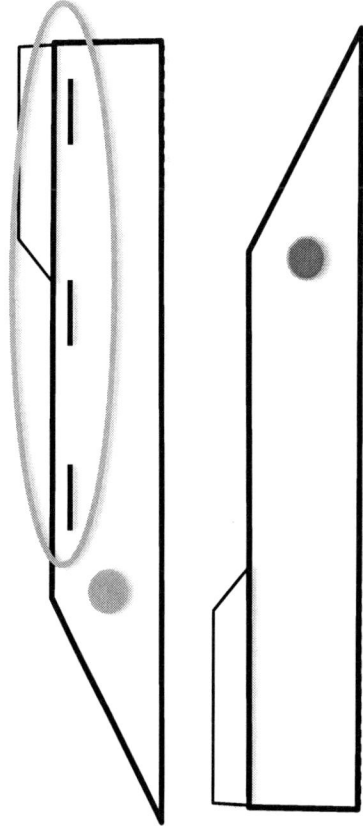

Side Views

# Operation 8
## (Optional; Not for Basic Game)
## Quality Check (for A4 size paper)

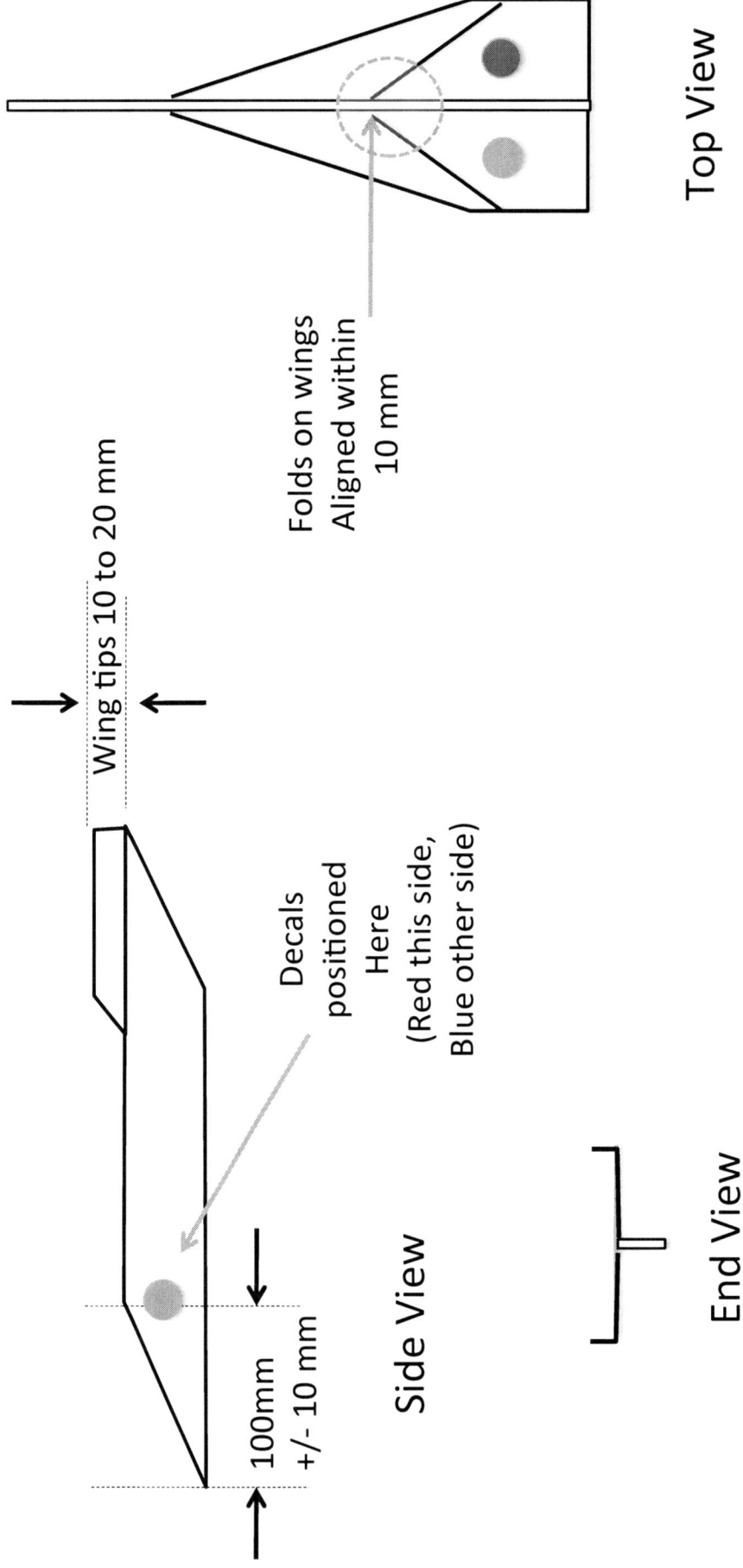

Wing tips 10 to 20 mm

Decals positioned Here
(Red this side, Blue other side)

100mm +/- 10 mm

Side View

End View

Folds on wings Aligned within 10 mm

Top View

# JIT Game Layout

The Lean Games and Simulations Book: Second Edition. Copyright © John Bicheno, 2014

# Customer Requirements

- The first delivery is due after 3 minutes

- At 2 minutes, and each minute thereafter, the Customer rolls two dice.

- The first Dice roll indicates how many planes will be required in the next minute

- The second Dice roll indicates the decal colours:
  - Roll 1 or 2: Decal colours: Red and Red
  - Roll 3 or 4: Decal colours Red and Blue (B on top right)
  - Roll 5 or 6: Decal colours Blue and Blue

- In other words, there is a 60 second order lead time.

- Backorders (i.e. late) not allowed.

- Stop after 12 minutes. (i.e. 10 deliveries)

# Dispatch Player Tasks

- You need to supply the Customer with his or her requirements.

- Keep a record of all orders received.

- Please keep a record of all orders that you are NOT able to meet. Record the number of airplanes not met.

- Part orders are acceptable.

- Late orders are NOT acceptable.

# Wing Decal Painting
## (for Player 5)

- Customer demand is known to be in the ratio of

  - One with two red decals

  - One with red and blue decals

  - One with two blue decals

- Therefore, for each batch of 6 airplanes you receive from the previous workstation, you should 'paint' two of each type given above.

# Wing Decal Painting
## (for Player 6)

- Customer demand is known to be in the ratio of

  – One with two red decals

  – One with red and blue decals

  – One with two blue decals

- Therefore, for each batch of 6 airplanes you receive from the previous workstation, you should 'paint' two of each type given above.

# Record Sheet: Round 1

| Minute | Orders Rec'd | Orders Met |
|--------|--------------|------------|
| 1 | 0 | 0 |
| 2 | | 0 |
| 3 | | |
| 4 | | |
| 5 | | |
| 6 | | |
| 7 | | |
| 8 | | |
| 9 | | |
| 10 | | |
| 11 | | |
| 12 | 0 | |

| Round Results | |
|---------------|---|
| No of players | |
| No NOT delivered | |
| Closing WIP & FGI | |
| Opening WIP & FGI | |
| Inventory buildup | |

Note: If Closing inventory is less than opening inventory, the difference must be added to the number of airplanes NOT delivered. This is for continuity reasons.

Note: Late orders are NOT accepted.

# Record Sheet: Round 2

| Minute | Orders Rec'd | Orders Met |
|--------|--------------|------------|
| 1 | 0 | 0 |
| 2 | | 0 |
| 3 | | |
| 4 | | |
| 5 | | |
| 6 | | |
| 7 | | |
| 8 | | |
| 9 | | |
| 10 | | |
| 11 | | |
| 12 | 0 | |

| Round Results | |
|---------------|---|
| No of players | |
| No NOT delivered | |
| Closing WIP & FGI | |
| Opening WIP & FGI | |
| Inventory buildup | |

Note: If Closing inventory is less than opening inventory, the difference must be added to the number of airplanes NOT delivered. This is for continuity reasons.

Note: Late orders are NOT accepted.

# Record Sheet: Round 3

| Minute | Orders Rec'd | Orders Met |
|--------|--------------|------------|
| 1 | 0 | 0 |
| 2 | | 0 |
| 3 | | |
| 4 | | |
| 5 | | |
| 6 | | |
| 7 | | |
| 8 | | |
| 9 | | |
| 10 | | |
| 11 | | |
| 12 | 0 | |

| Round Results | |
|---------------|---|
| No of players | |
| No NOT delivered | |
| Closing WIP & FGI | |
| Opening WIP & FGI | |
| Inventory buildup | |

Note: If Closing inventory is less than opening inventory, the difference must be added to the number of airplanes NOT delivered. This is for continuity reasons.

Note: Late orders are NOT accepted.

# Airplane Lean Game Measures

- Number of airplanes NOT delivered

- Total WIP and finished goods at end of game

- Number of rejects due to quality reasons

- Number of players

- Note: Final (ending) inventory must be at least as much as initial (beginning) inventory, in order to assure continuity. Inventory here means WIP and finished goods. If ending inventory is less than starting inventory, the number will be deducted from total airplanes delivered.

# Quality Checks

- Your task is to do Quality spot checks, according to the specification

- You work for the Paper Airplane company, but have responsibilities for airplane safety to the CAA (Cardboard Aircraft Authority)

- So, you should not pass on defective airplanes.

- You do not need to measure every plane . Measure a few planes but visual checks are Ok.

- If an airplane is defective it must be scrapped. No rework.

# Player Instructions – Round 1

- Collect the raw material or partly built up airplanes from the previous operation in batches of 6.

- Undertake your required step. (Use the Airplane build instructions.)

- Place the part-made or complete airplanes in the outbound work in process area.

# Improvement Activity

- Identify and remove waste
- Map the process
- Implement your changes
- You may change anything, except the basic airplane design.
- The Customer requirement mix will not change and cannot be changed.
- You should start again with zero inventory.
- Your instructor will indicate how much time you have to make and implement changes.

# JIT Lean Game: Possible Changes

- Layout
- One piece flow
- Pull: Kanban / DBR / CONWIP
- Postponement
- Successive and own quality checks
- Takt time (practical) 60/3.5 = 17.1 seconds. Say 15 seconds.
- Line (activity) balance
- Reallocation of resources.
- Visual management
- Problem solving
- Pokayoke (?)

# Layout Types Game

In 12 minutes, build and inspect airplanes according to each of the following types:

- Everyone builds and inspects the complete airplane individually

- Two teams: first team builds half the airplane individually (in parallel); second team builds the second half individually. WIP area between.

- Two parallel assembly lines, with job specialization. In a U shape cell with X working?

- One long assembly line, with job specialization.

The Lean Games and Simulations Book: Second Edition. Copyright © John Bicheno, 2014

# Cell Layout Types Airplane Game

## Table of Considerations

| Considerations V \ Types > | Short-cycle Assembly Line Traditional and CONWIP | Individuals do the whole task | Two parallel lines, each with greater work content | Two groups. Each player makes half. WIP in between groups. |
|---|---|---|---|---|
| Productivity | | | | |
| Training | | | | |
| Inventory | | | | |
| Visibility | | | | |
| Traceability | | | | |
| Boredom | | | | |
| Quality | | | | |
| Tooling | | | | |
| Supervision | | | | |

The Lean Games and Simulations Book: Second Edition. Copyright © John Bicheno, 2014

# Cell Layout Types Airplane Game : The Four Types

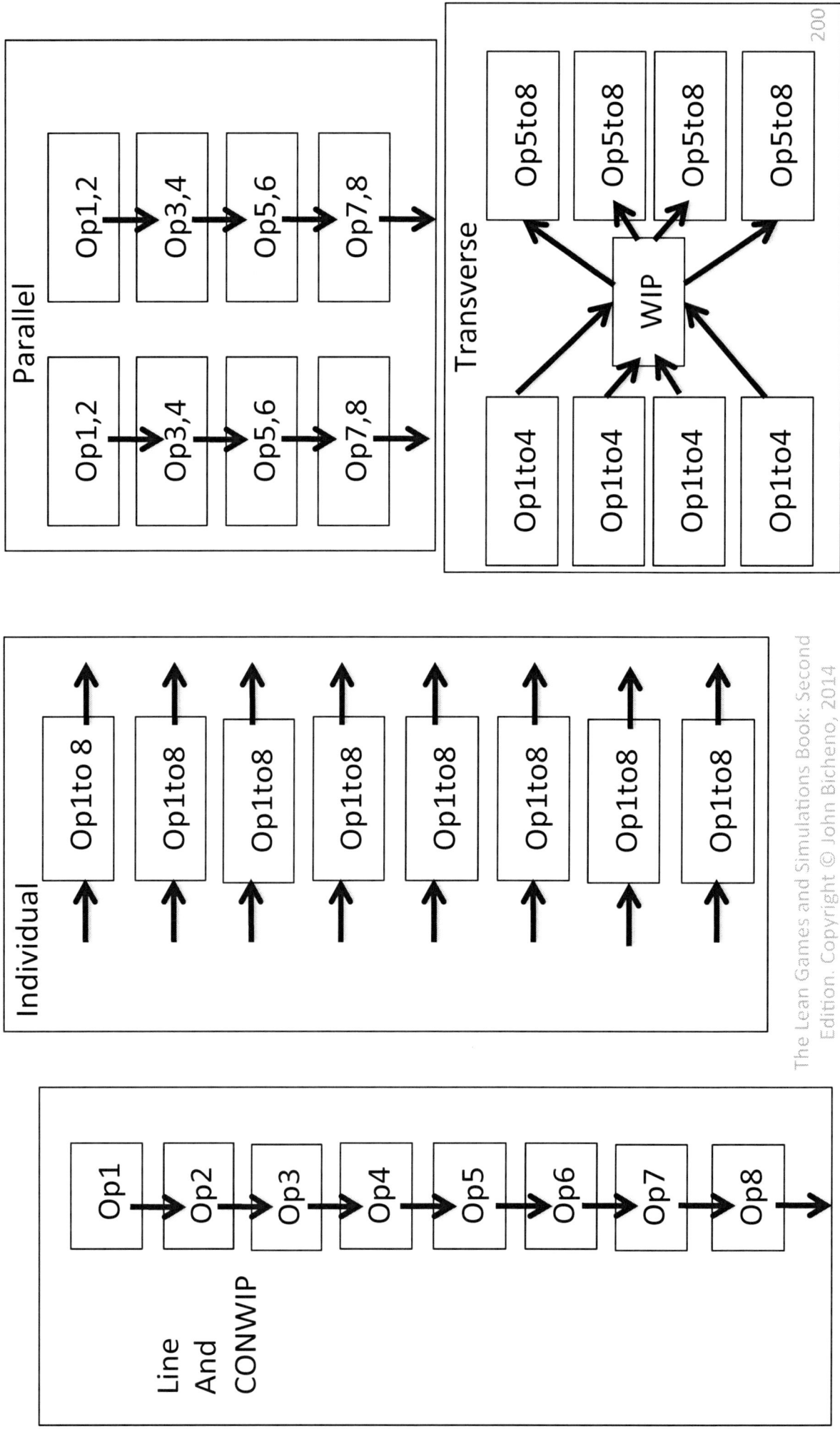

**Parallel**

| Op1,2 | Op3,4 | Op5,6 | Op7,8 |

| Op1,2 | Op3,4 | Op5,6 | Op7,8 |

**Transverse**

Op1to4 → WIP → Op5to8 (×4)

**Individual**

Op1to 8, Op1to8, Op1to8, Op1to8, Op1to8, Op1to8, Op1to8, Op1to8

**Line And CONWIP**

Op1 → Op2 → Op3 → Op4 → Op5 → Op6 → Op7 → Op8

# 'Bucket Brigade' Line Balancing

## Game Sheet

Round 1 : Conventional Line Balancing

Round 2 : Bucket Brigade Balance

## Initial Operation Timings

| Op → | Times in seconds for 5 planes | | | | |
|---|---|---|---|---|---|
| | 1 | 2 | 3 | 4 | 5 |
| 1 | | | | | |
| 2 | | | | | |
| 3 | | | | | |
| 4 | | | | | |
| 5 | | | | | |
| 6 | | | | | |
| 7 | | | | | |

## Results: Planes made in 12 minutes

| | Round 1 | Round 2 |
|---|---|---|
| WIP | | |
| Throughput | | |
| Rejects | | |

# Job Breakdown Sheet Template

| Important Steps | Key Points | Reasons |
|---|---|---|
| (Something that advances the work.) | Anything that may<br>1.Make or break the job<br>2.Injure the worker<br>3.Make the work easier. 'Knacks' or 'Tricks'. | Reasons for the Key points |
| 1. | | |
| 2. | | |
| 3. | | |
| 4. | | |

After Patrick Graup and Robert Wrona, *The TWI Workbook*, Productivity

# Squares Games

Square 1 : Front

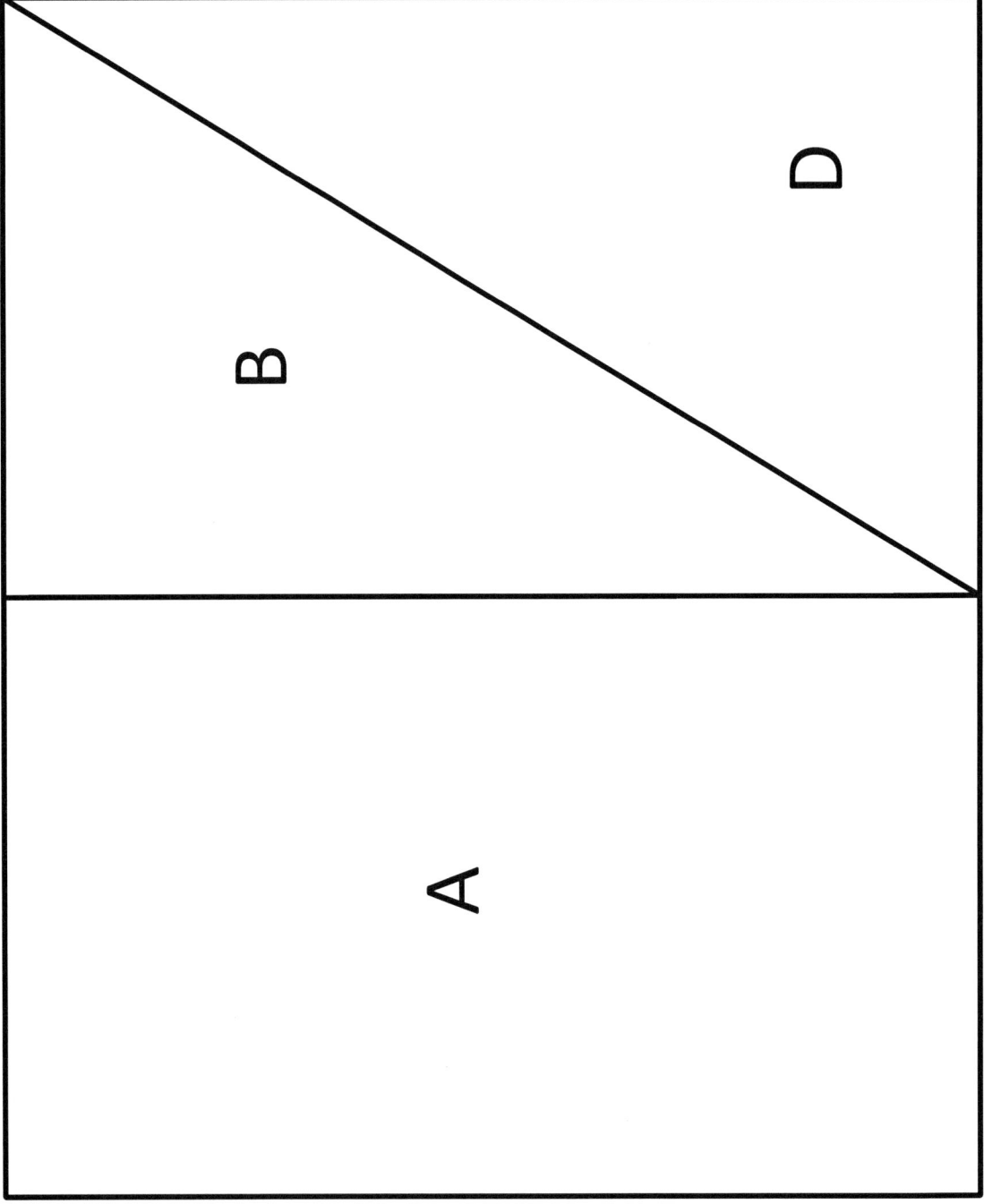

A

B

D

Square 1 : Back

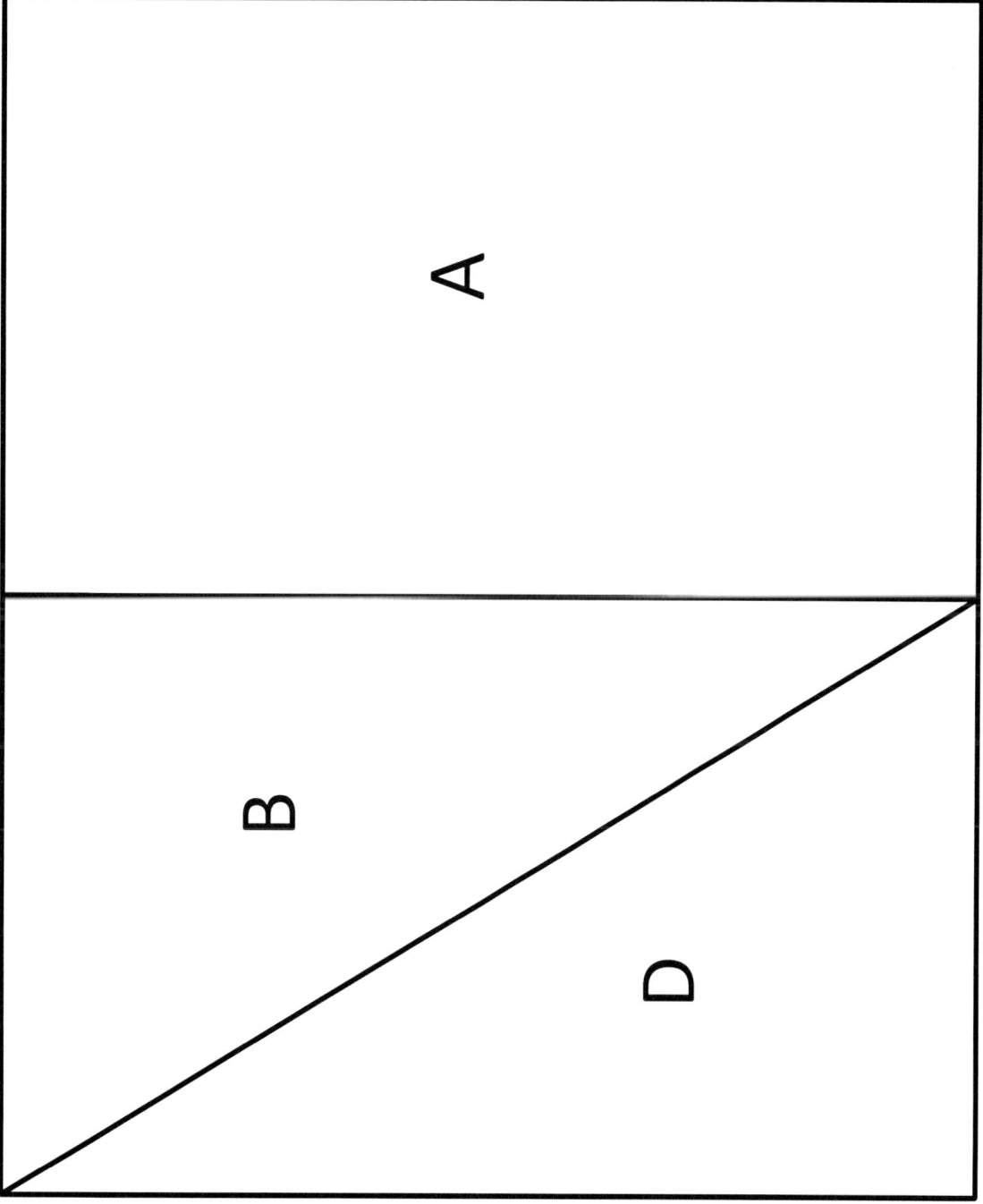

Square 2 : Front

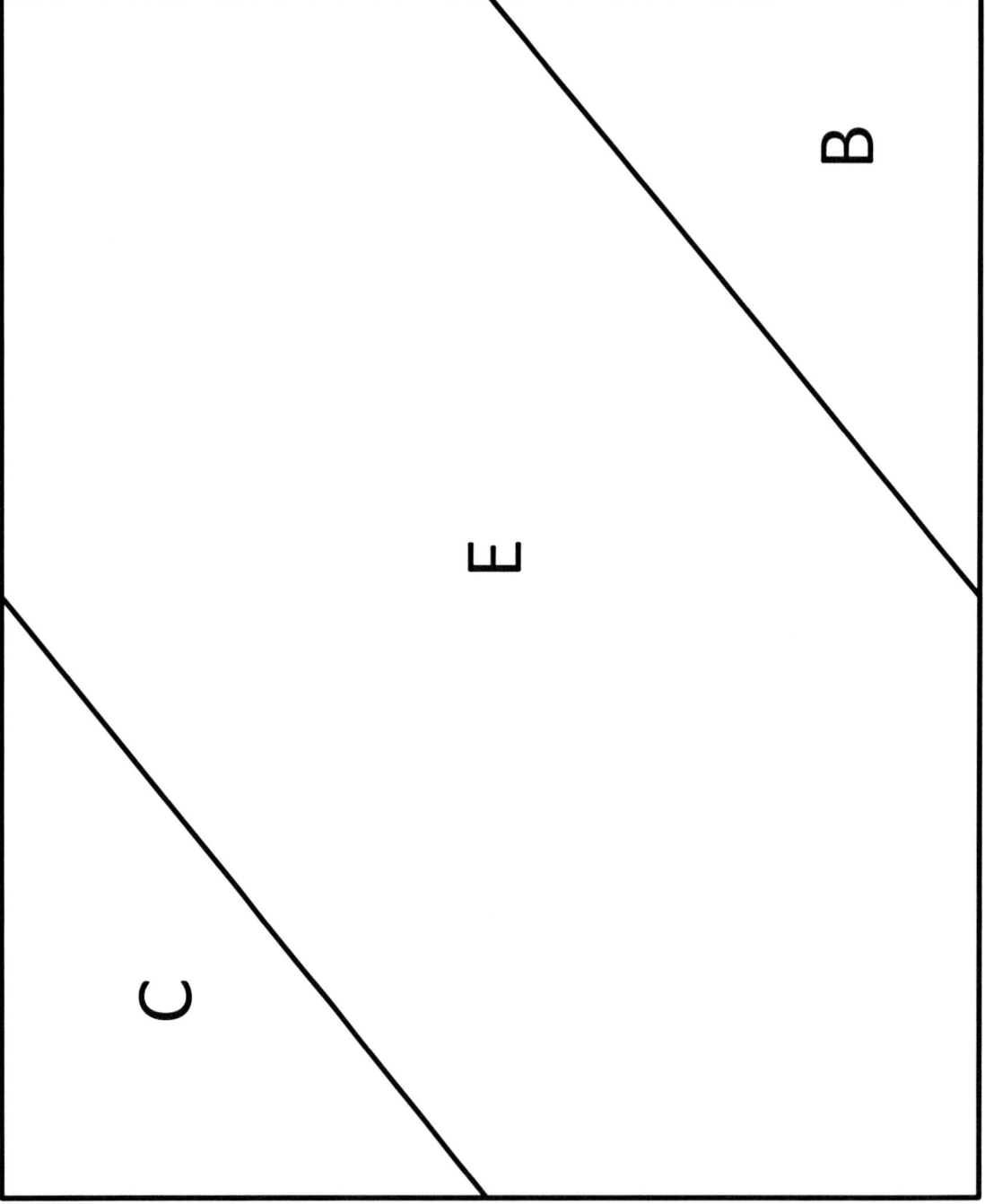

C

E

B

Square 2 : Back

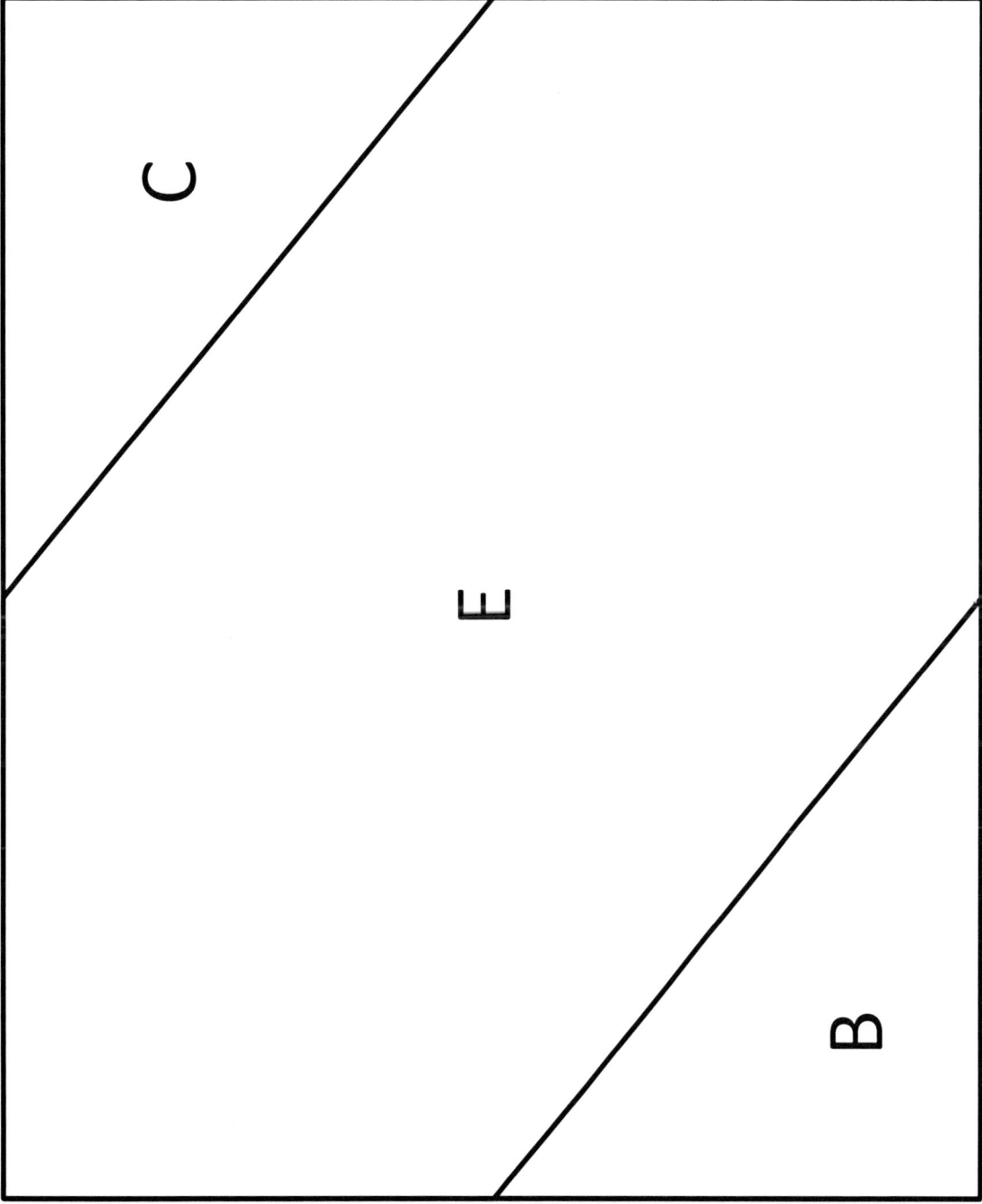

Square 3 : Front

Square 3 : Back

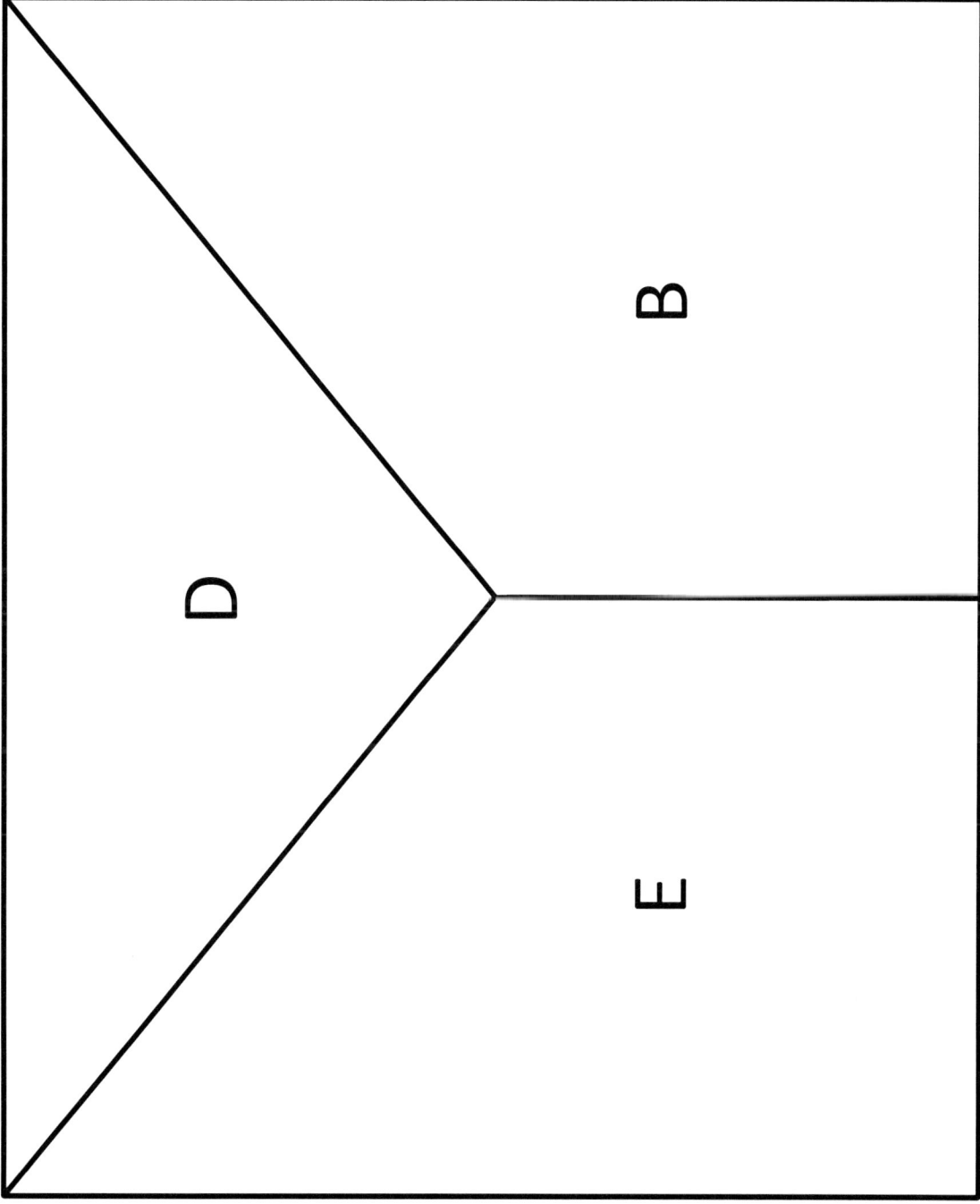

Square 4 : Front

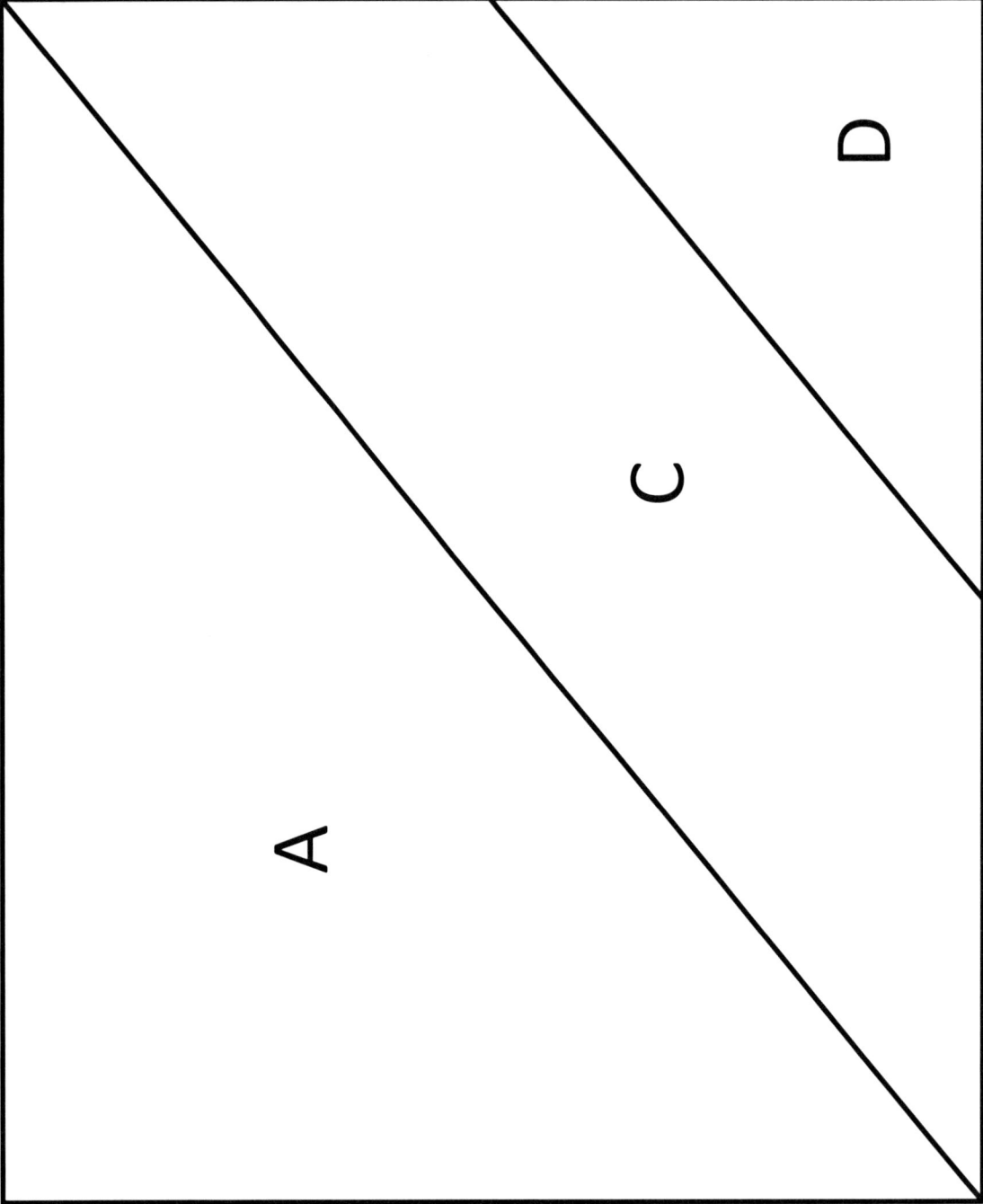

A

C

D

Square 4 : Back

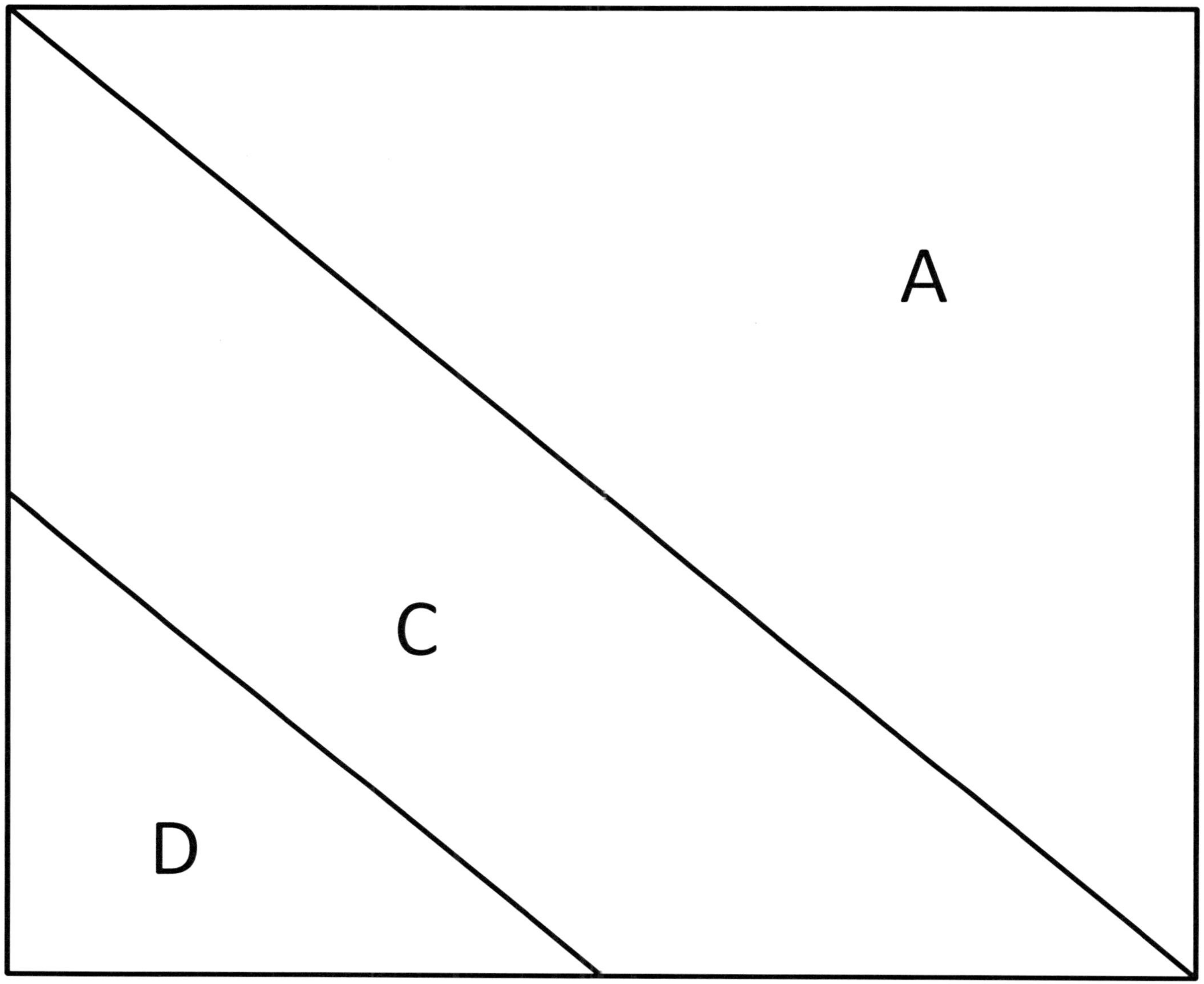

A

C

D

Square 5 : Front

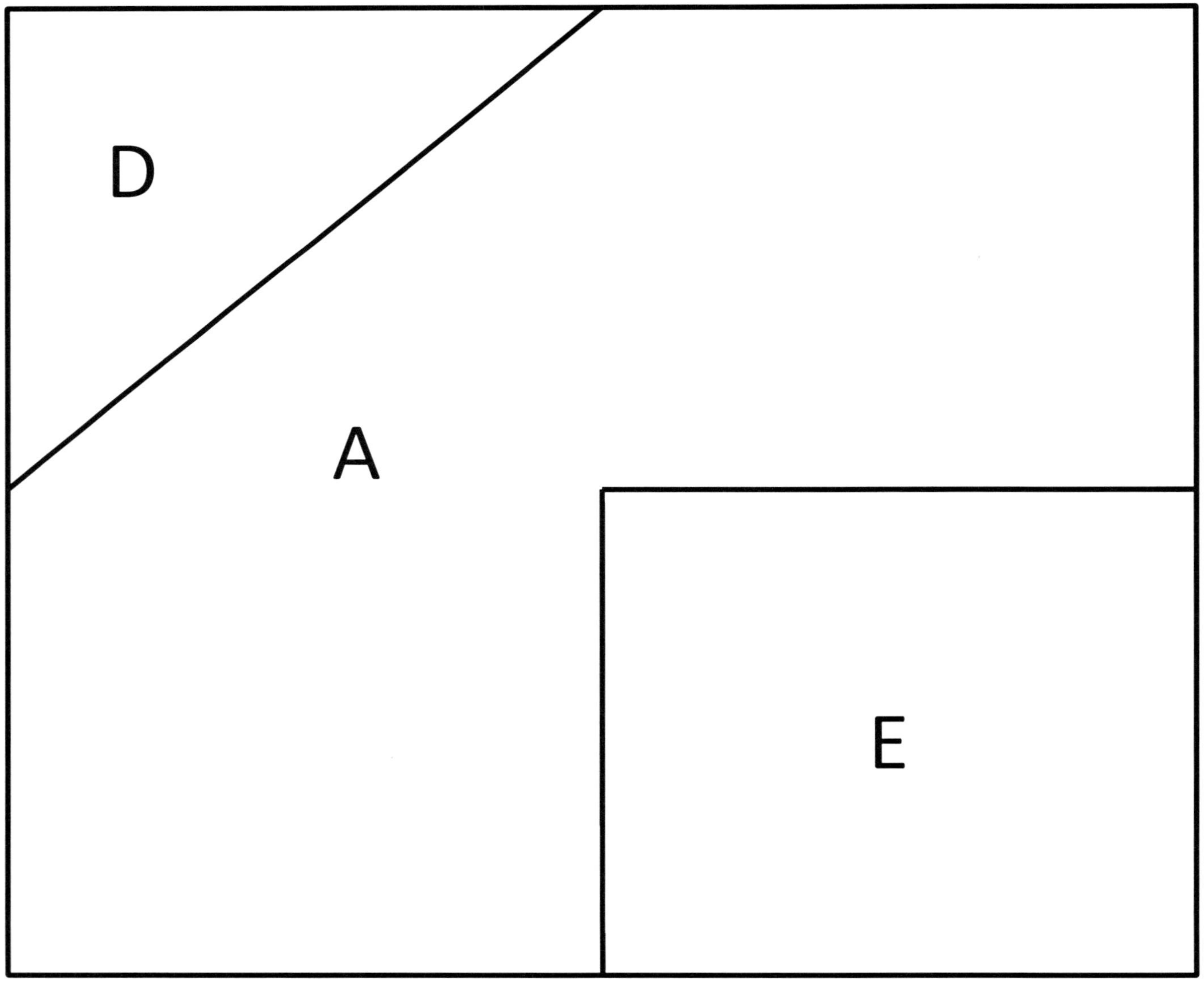

D

A

E

Square 1 : Back

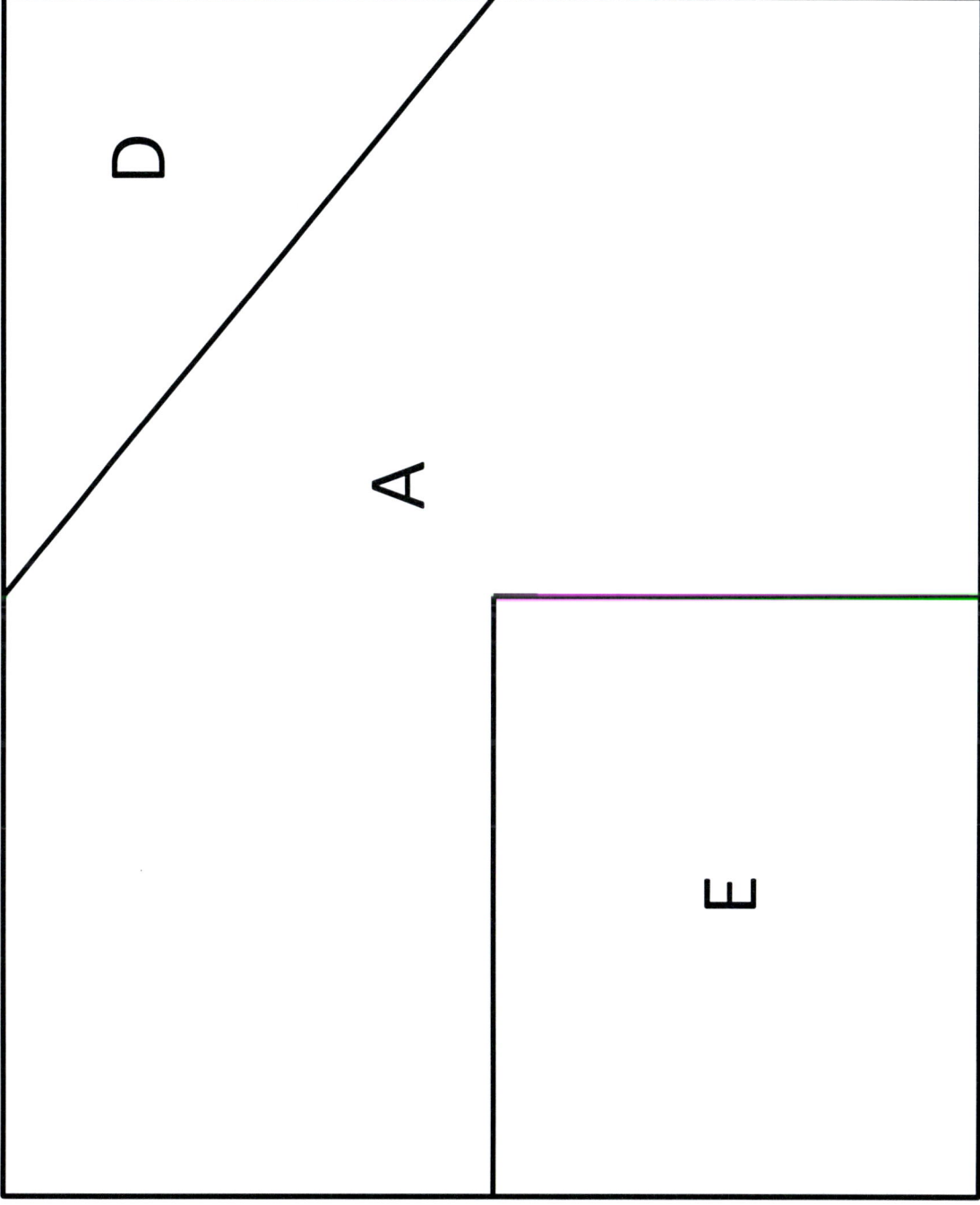

D

A

E

Extras: For 5S Game

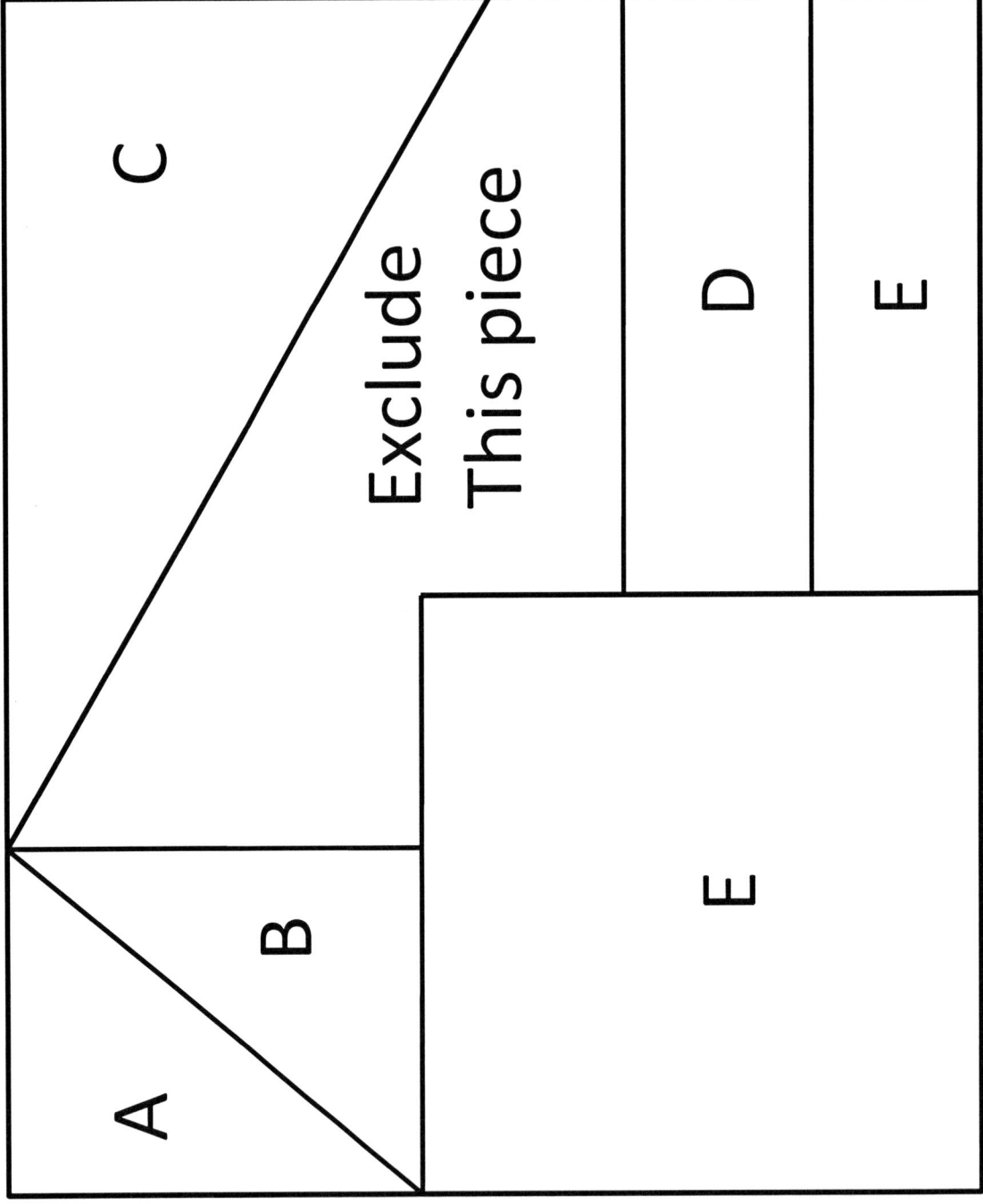

A

B

C

D

E

E

Exclude
This piece

Extras: For 5S Game

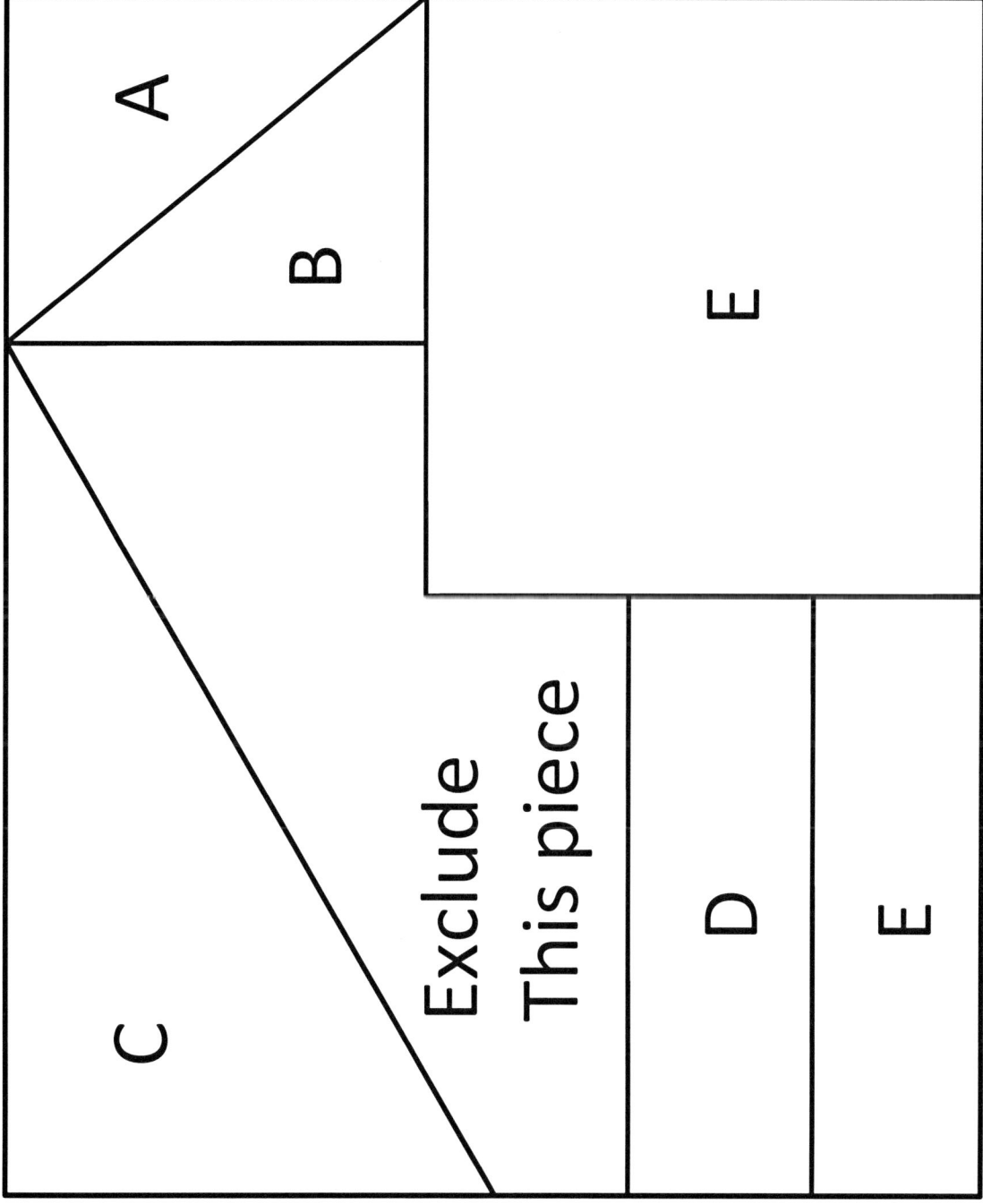

A

B

C

Exclude
This piece

D

E

E

Extras: For 5S Game

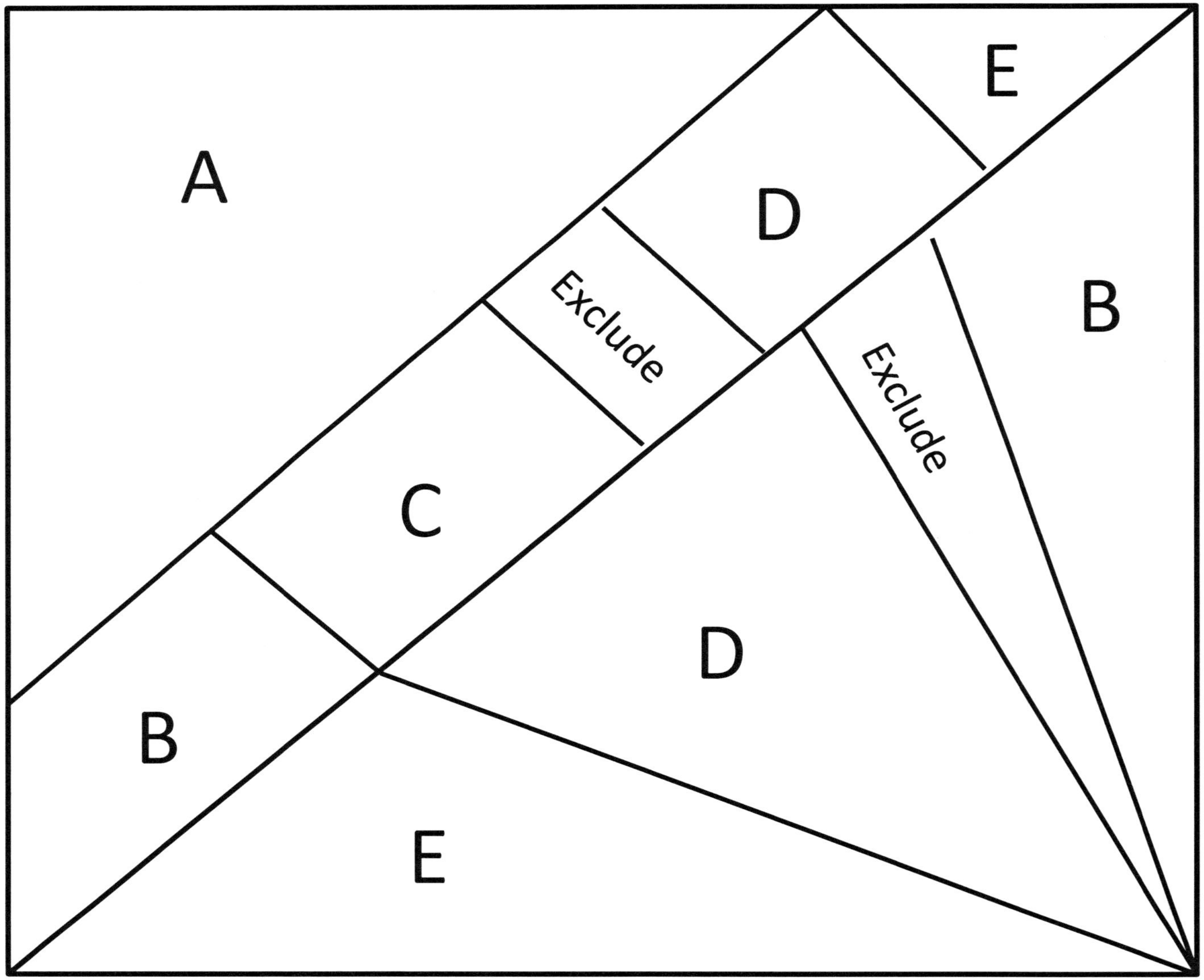

A
E
D
Exclude
C
B
Exclude
D
B
E


The Lean Games and Simulations Book: Second Edition. Copyright © John Bicheno, 2014

Extras: For 5S Game

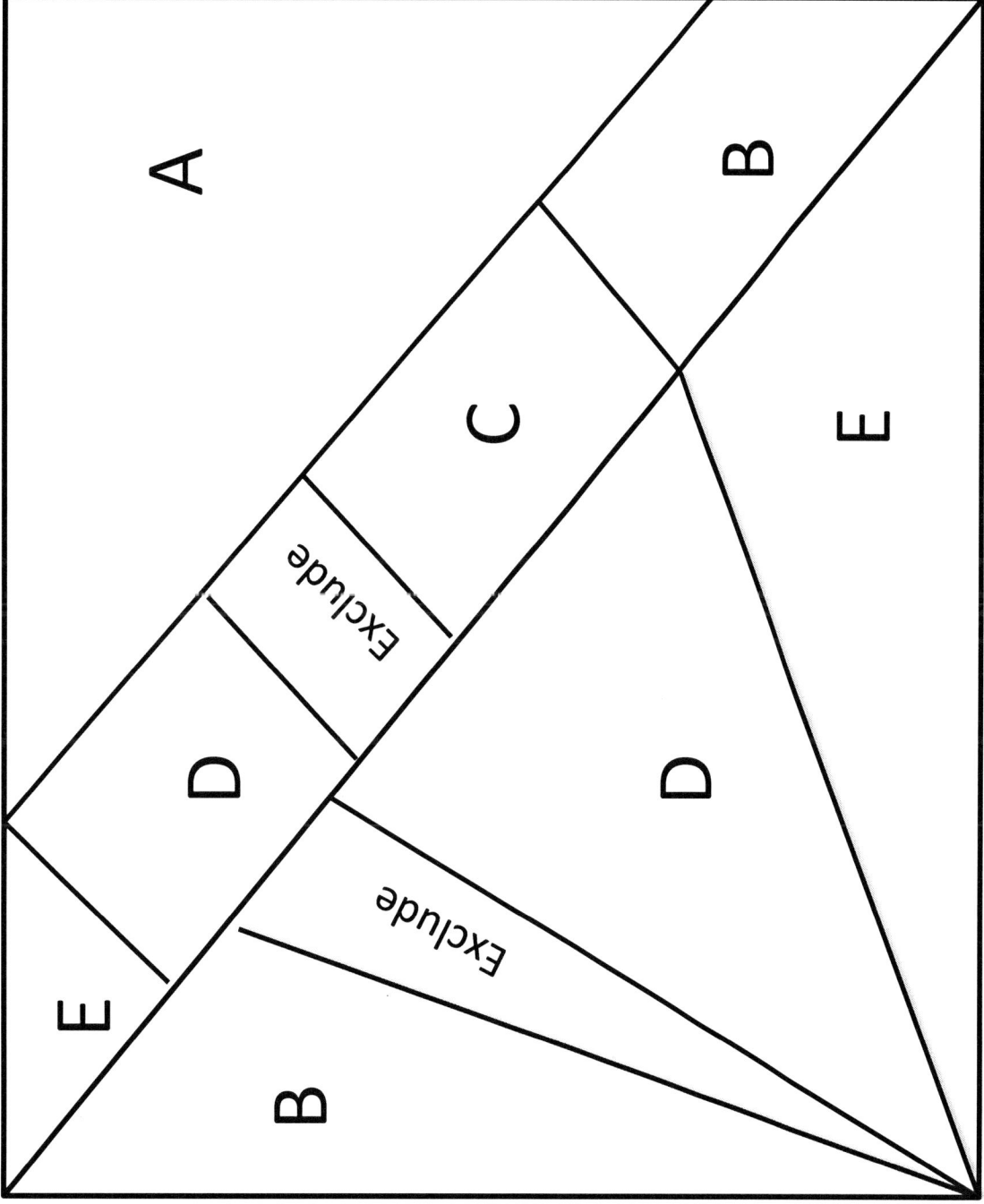

A

B

C

Exclude

E

D

D

Exclude

E

B

Extras: For Changeover Game: Front.

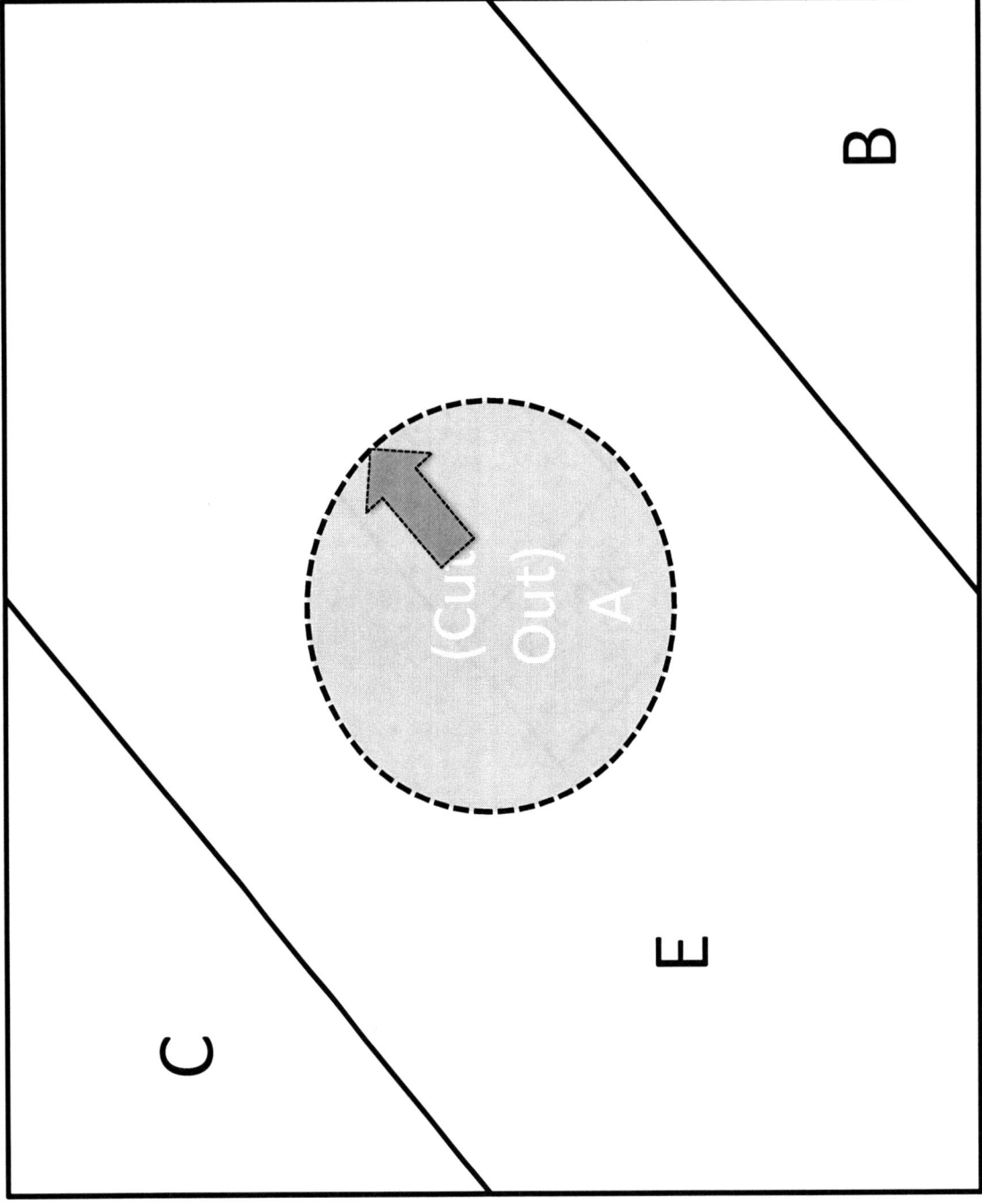

B

C

E

(Cut
Out)

A

Extras: For Changeover Game: Back.

C

E

B

A
(Cut Out)

# Changeover Instructions

- Assemble into a rectangle containing 6 squares. The top right square of the rectangle must be the 'Heat Treated' square with the circle inserted.

- The circle requires heating for 20 seconds before insertion in the hole. Once heated, it can only be handled with two pens / pencils.

- The arrow on the circle must be aligned with a line drawn from bottom left corner of inner rectangle to top right corner of inner rectangle.

- During changeover reduction, sticky tape or staples may not be used.

The Lean Games and Simulations Book: Second Edition. Copyright © John Bicheno, 2014

# Preparation

- Cut out pieces and place in an envelope. Paper is OK, but cardboard is better

- The circle must be placed in another envelope, located (say) 5m away from the first envelope.

- The diagonal line, for alignment with the arrow, must not be drawn on the sheet.

- The heater for the circle is shown on a separate sheet.

The Lean Games and Simulations Book: Second Edition. Copyright © John Bicheno, 2014

# Heater

HOT

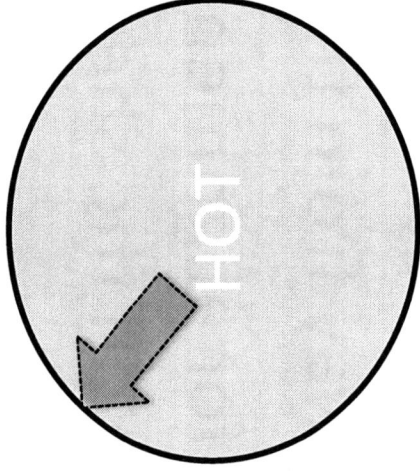

Place the circle in this heater for 20 seconds prior to installation.
Handle the circle with two pencils, when 'hot'.
Align the arrow with the diagonal, drawn corner to corner.

The Circle will remain at an acceptable temperature for 20 seconds after heating.

The Heater Square containing the circle may not be moved once heat treatment has begun.

# Dice Games

# Queue Time, Variation and Utilization

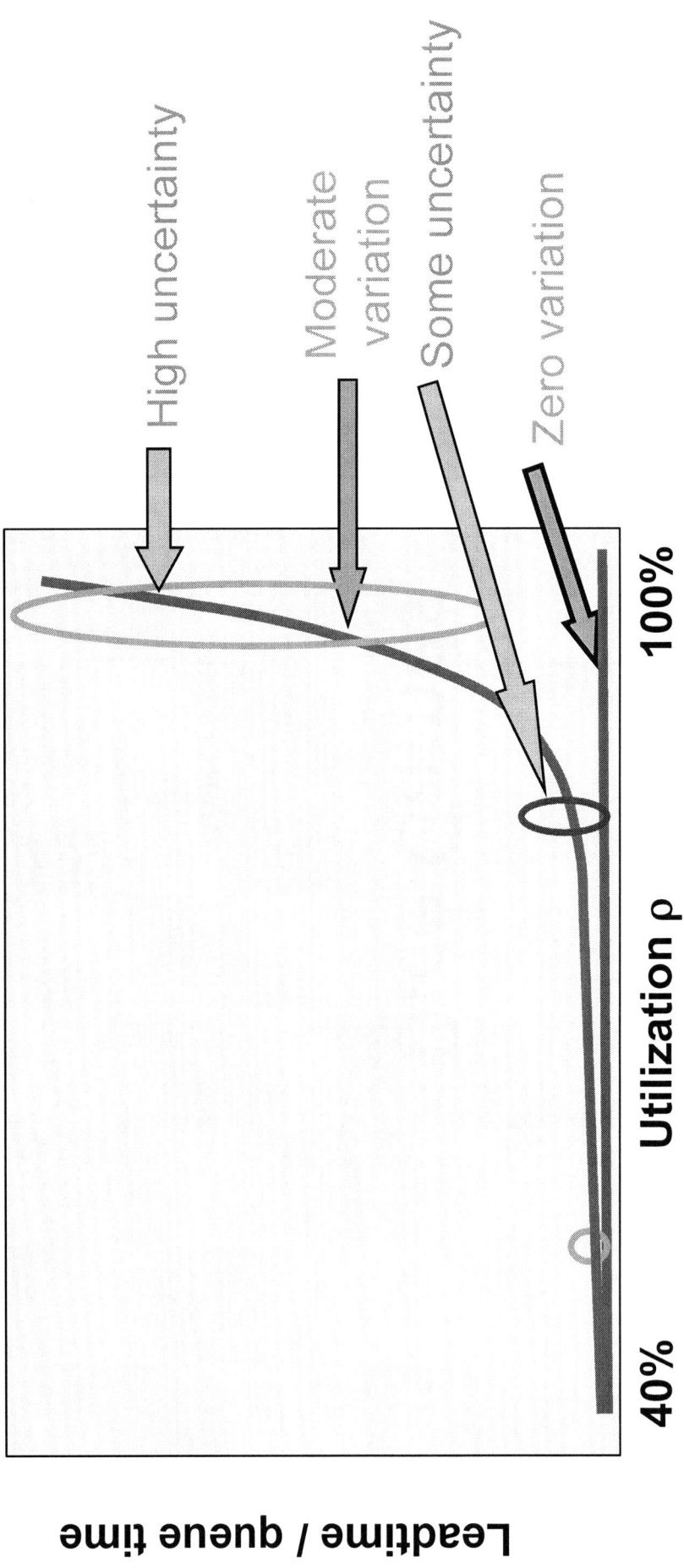

**Leadtime / queue time**

High uncertainty

Moderate variation

Some uncertainty

Zero variation

40%    **Utilization ρ**    100%

Utilization is related to Muri (Overload)

Variation is Mura (Two Types: Arrival (Demand) and Process)

# Muda, Muri, Mura

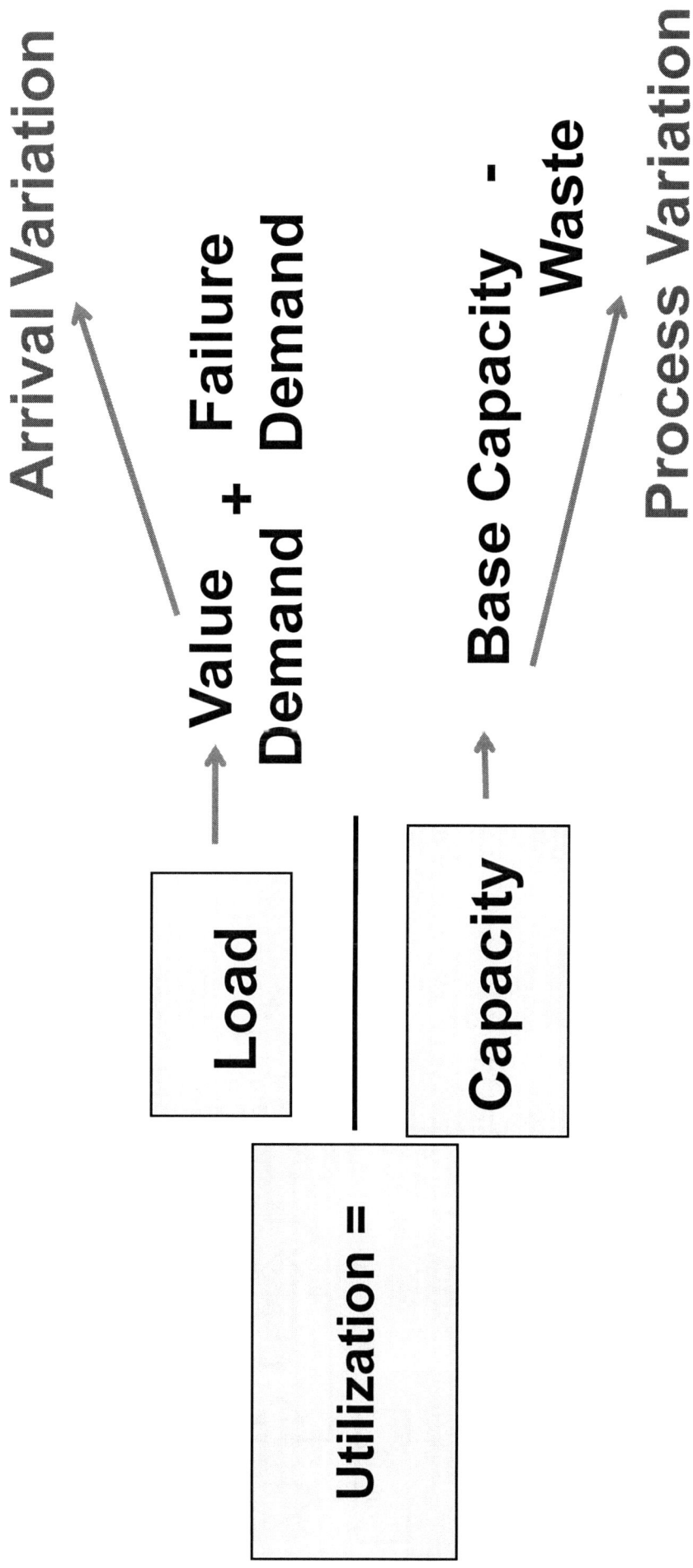

$$\text{Utilization} = \frac{\text{Load}}{\text{Capacity}}$$

Load → Value Demand + Failure Demand → Arrival Variation

Capacity → Base Capacity − Waste → Process Variation

# Drum Buffer Rope with Priority

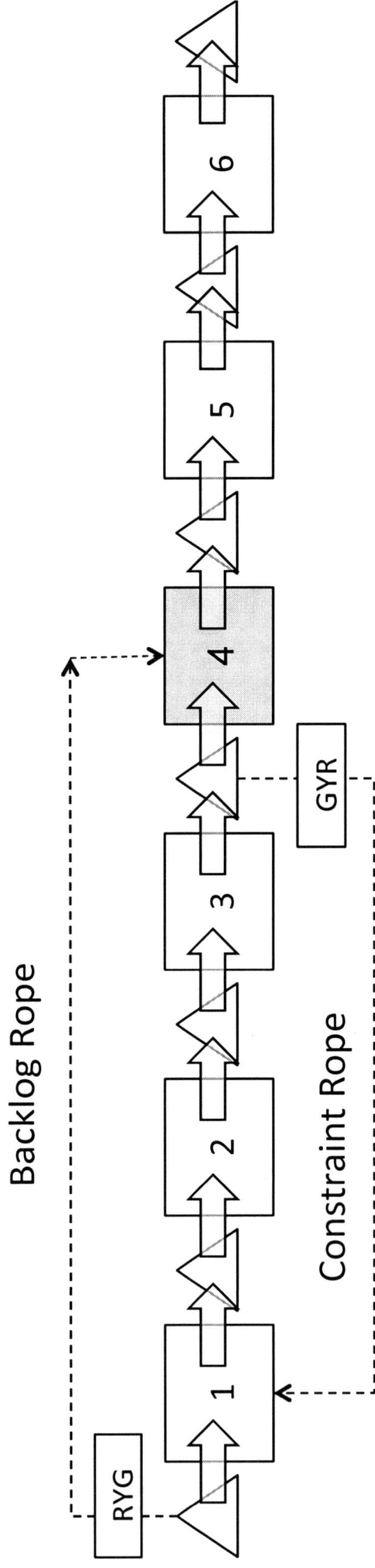

Backlog Rope

Constraint Rope

RYG

GYR

The Lean Games and Simulations Book: Second Edition. Copyright © John Bicheno, 2014

# Drum Buffer Rope with Priority:
## FIFO Lanes

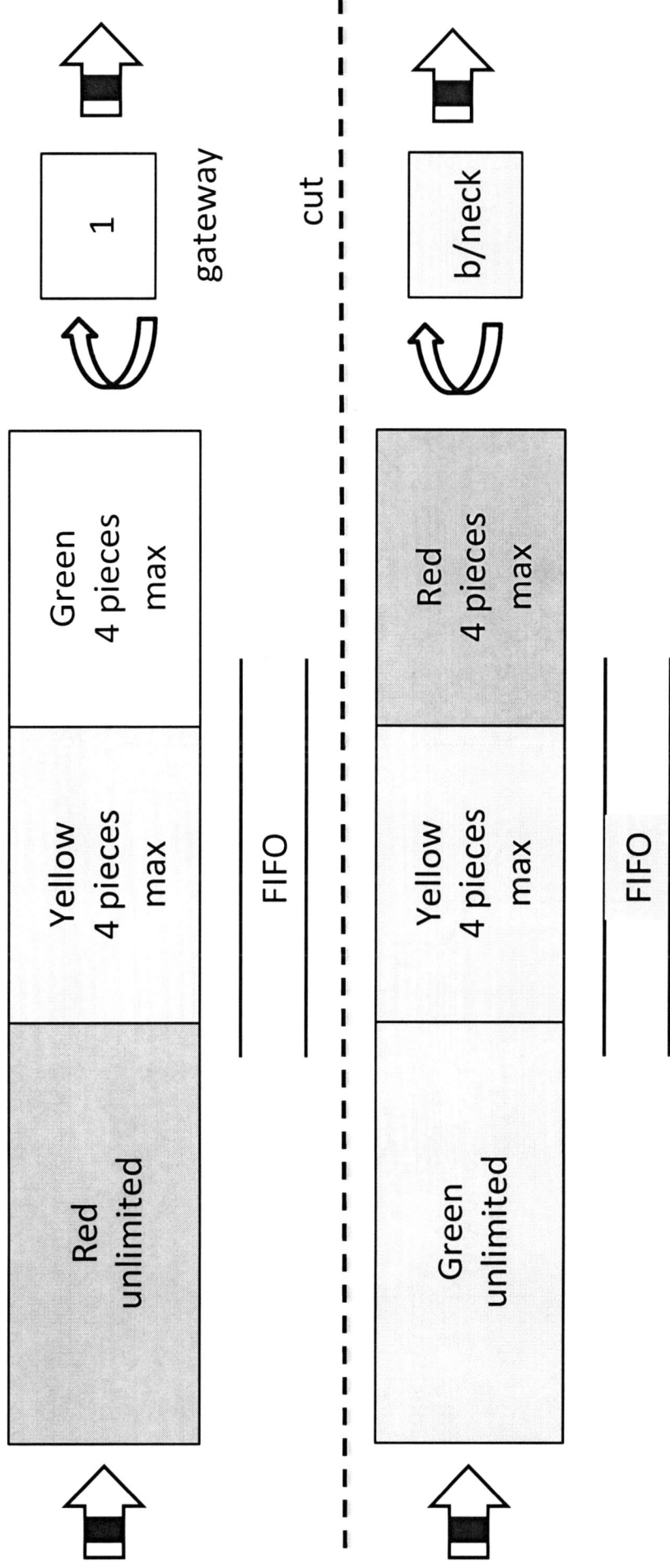

| Red unlimited | Yellow 4 pieces max | Green 4 pieces max |
|---|---|---|

FIFO

gateway → 1

cut

| Green unlimited | Yellow 4 pieces max | Red 4 pieces max |
|---|---|---|

FIFO

b/neck

The Lean Games and Simulations Book: Second Edition. Copyright © John Bicheno, 2014

# Single Piece Flow Dice Game (1)

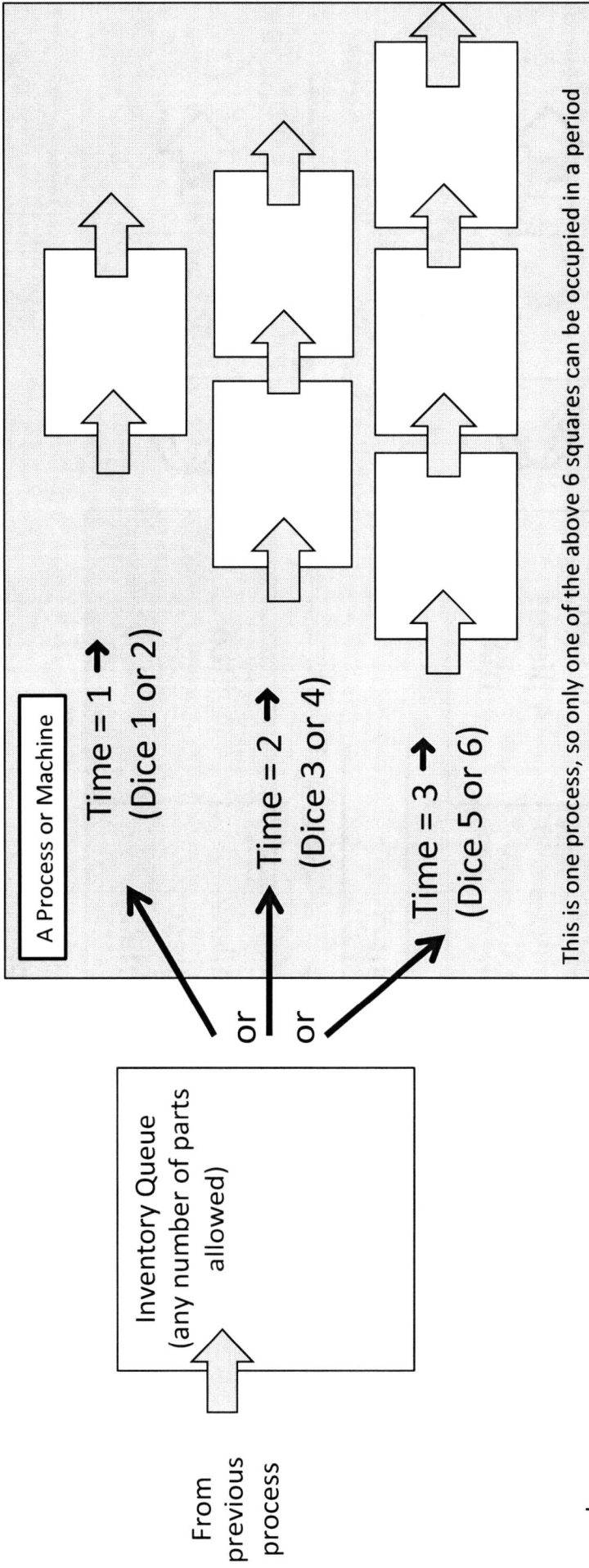

From previous process

Inventory Queue
(any number of parts allowed)

or

or

Time = 1 →
(Dice 1 or 2)

Time = 2 →
(Dice 3 or 4)

Time = 3 →
(Dice 5 or 6)

A Process or Machine

This is one process, so only one of the above 6 squares can be occupied in a period

Next period → Move any in-process part → Parts still in process? → N → Inventory in Queue? → Y → Roll dice → Place part into process

Wait ← Y (Parts still in process)

N (Inventory in Queue?) 

Wait

## Remember

- Only one part at a time in the process
- A dice is <u>not</u> rolled every period
- A dice is only rolled when a new part is started
- To start a new part, the process must be empty and inventory must be available.
- Wait for the instructor to signal a new period.
- Please don't jump steps.

# Single Piece Flow Dice Game (2)

(Reduced Variation;
Ave process time = 2)

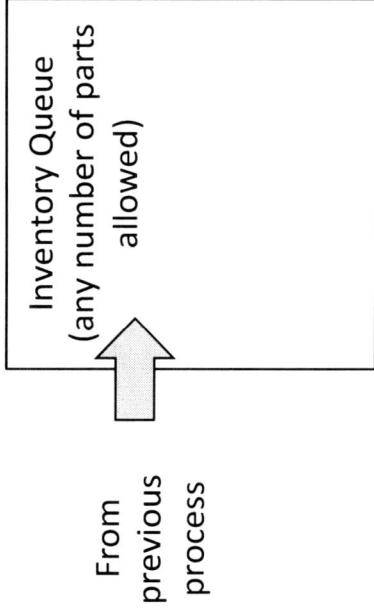

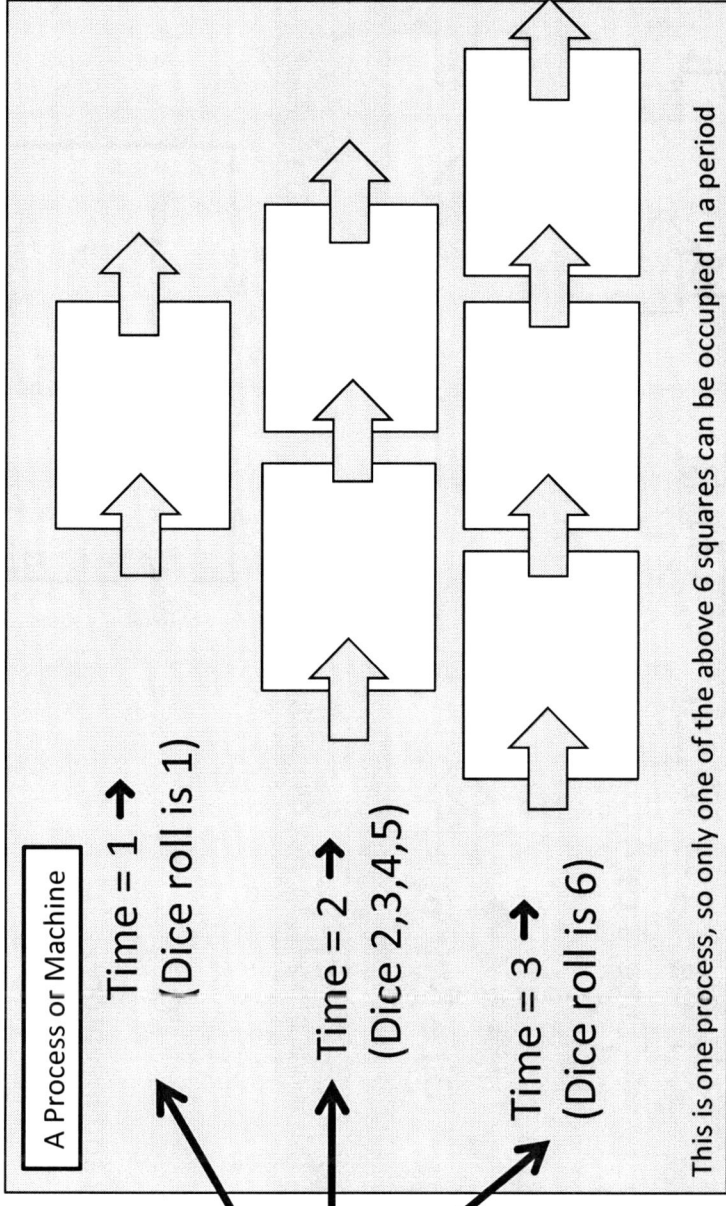

From
previous
process

Inventory Queue
(any number of parts
allowed)

A Process or Machine

Time = 1 ↑
(Dice roll is 1)

or

Time = 2 ↑
(Dice 2,3,4,5)

or

Time = 3 ↑
(Dice roll is 6)

This is one process, so only one of the above 6 squares can be occupied in a period

→ Next period

Wait ┐

Move any in-process part

Y ┘

N ┐

Parts still in process?

N ┐

Inventory in Queue?

Y ┘

Roll dice

Place part into process

Wait

Remember

- Only one part at a time in the process
- A dice is <u>not</u> rolled every period
- A dice is only rolled when a new part is started
- To start a new part, the process must be empty and inventory must be available.
- Wait for the instructor to signal a new period.
- Please don't jump steps.

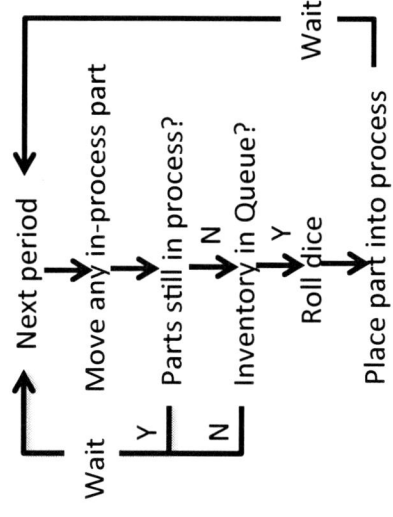

# Single Piece Flow Dice Game (3)

(Process ave. time = 3)

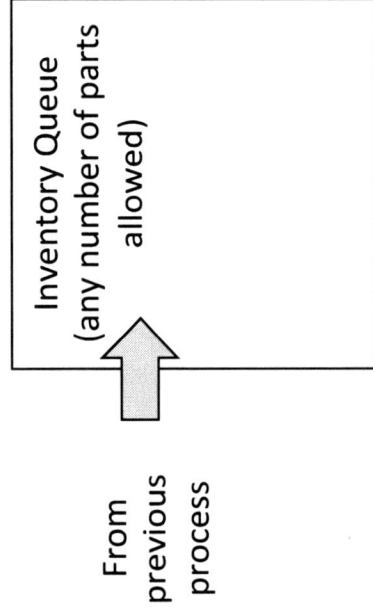

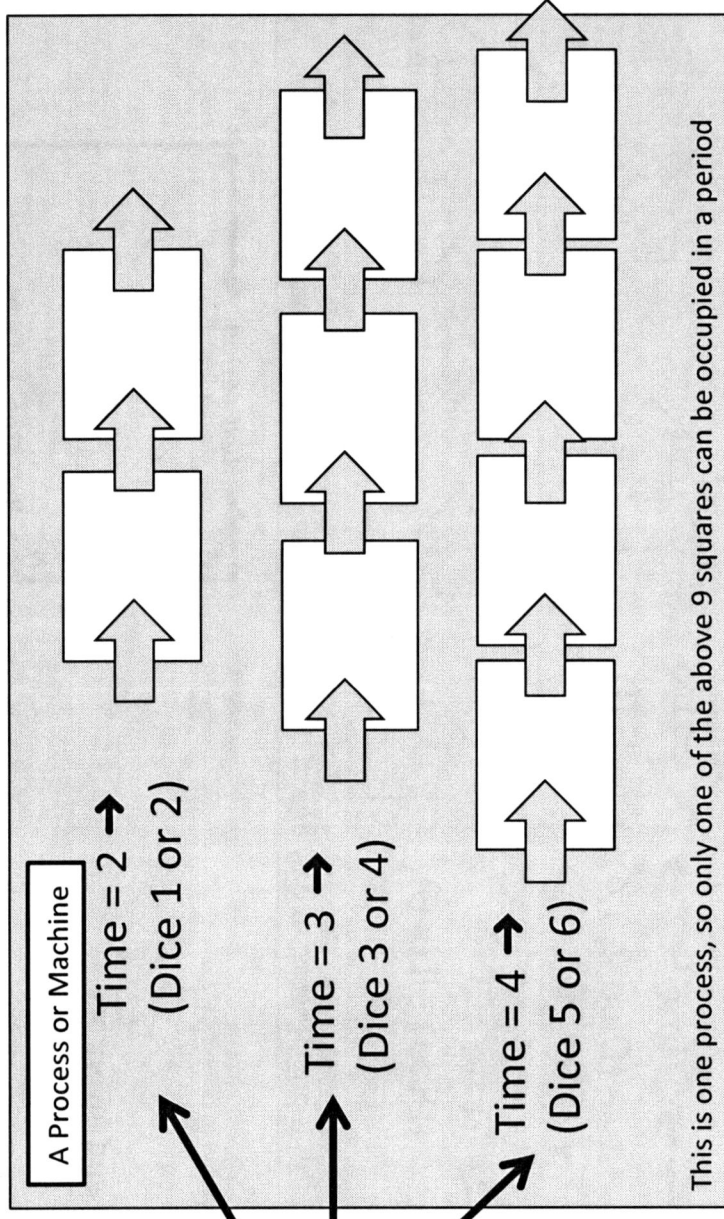

A Process or Machine

Time = 2 ↑
(Dice 1 or 2)

Time = 3 ↑
(Dice 3 or 4)

Time = 4 ↑
(Dice 5 or 6)

Inventory Queue
(any number of parts allowed)

From previous process

or

or

This is one process, so only one of the above 9 squares can be occupied in a period

Next period → Move any in-process part → Parts still in process?

Y → Wait

N → Inventory in Queue?

Y → Roll dice → Place part into process

N → Wait

**Remember**

- Only one part at a time in the process
- A dice is <u>not</u> rolled every period
- A dice is only rolled when a new part is started
- To start a new part, the process must be empty and inventory must be available.
- Wait for the instructor to signal a new period.
- Please don't jump steps.

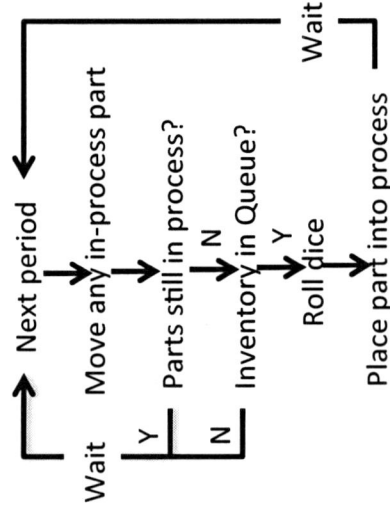

# Single Piece Flow Dice Game (4)

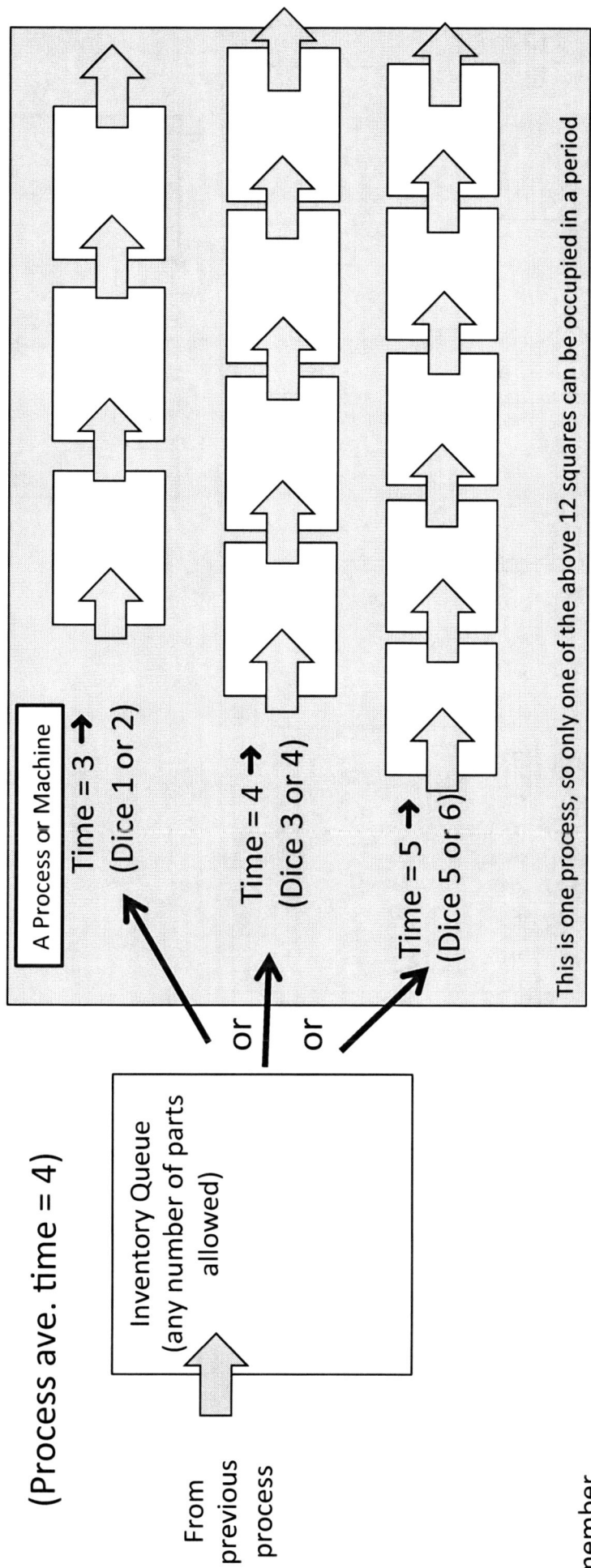

(Process ave. time = 4)

A Process or Machine

Time = 3 → (Dice 1 or 2)

Time = 4 → (Dice 3 or 4)

Time = 5 → (Dice 5 or 6)

This is one process, so only one of the above 12 squares can be occupied in a period

Inventory Queue (any number of parts allowed)

From previous process

Next period

Move any in-process part

Parts still in process?

Y — Wait

N

Inventory in Queue?

Y — Roll dice → Place part into process

N — Wait

**Remember**

- Only one part at a time in the process
- A dice is _not_ rolled every period
- A dice is only rolled when a new part is started
- To start a new part, the process must be empty and inventory must be available.
- Wait for the instructor to signal a new period.
- Please don't jump steps.

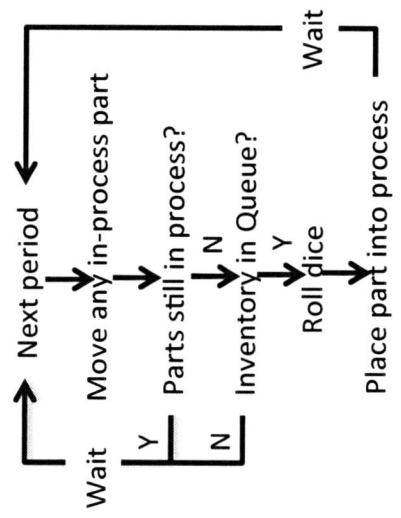

The Lean Games and Simulations Book: Second Edition. Copyright © John Bicheno, 2014

# Single Piece Flow Dice Game : Batch

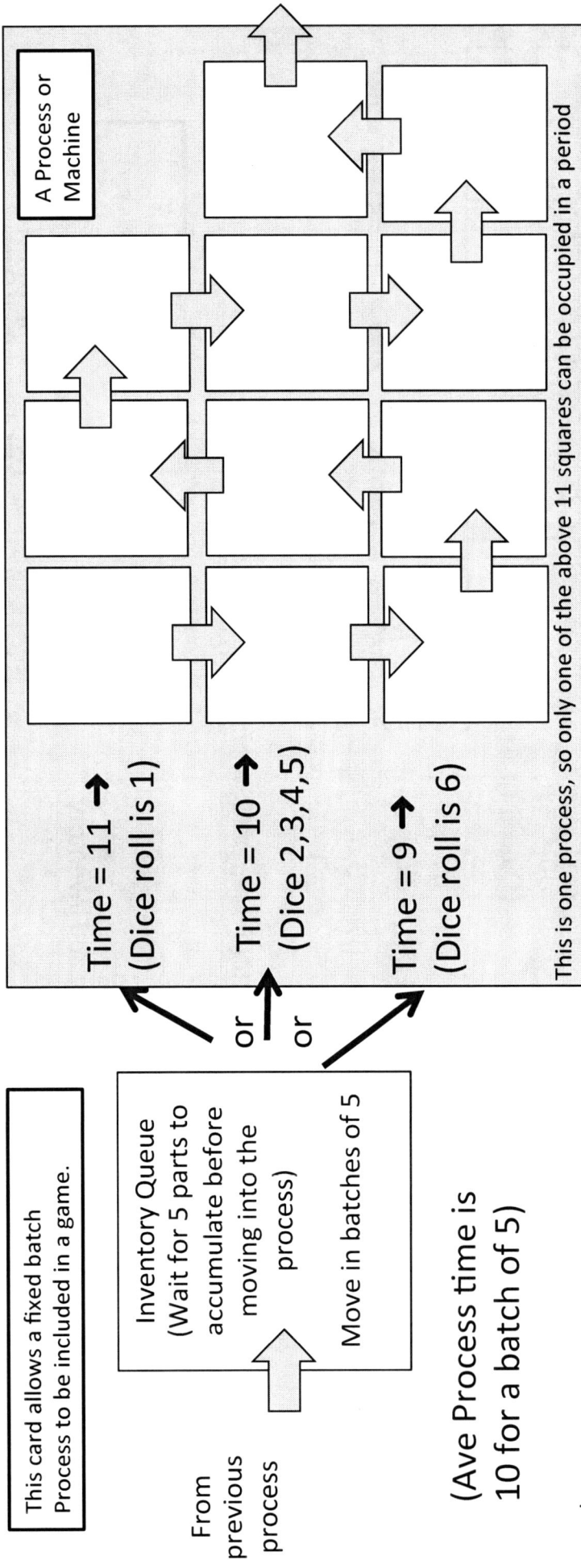

This card allows a fixed batch Process to be included in a game.

A Process or Machine

Time = 11 → (Dice roll is 1)

or

Time = 10 → (Dice 2,3,4,5)

or

Time = 9 → (Dice roll is 6)

This is one process, so only one of the above 11 squares can be occupied in a period

From previous process

Inventory Queue (Wait for 5 parts to accumulate before moving into the process)

Move in batches of 5

(Ave Process time is 10 for a batch of 5)

Remember

- Only one part at a time in the process
- A dice is <u>not</u> rolled every period
- A dice is only rolled when a new part is started
- To start a new part, the process must be empty and inventory must be available.
- Wait for the instructor to signal a new period.
- Please don't jump steps.

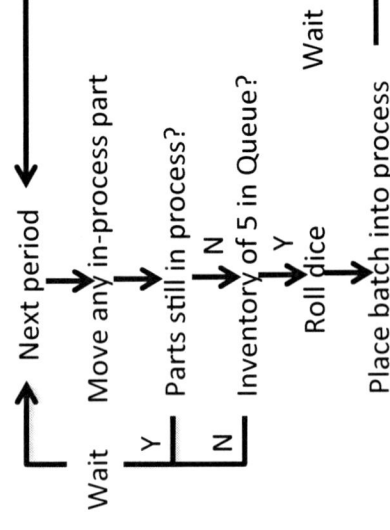

Next period → Move any in-process part → Parts still in process?

Wait

Y → Wait

N → Inventory of 5 in Queue? — Y → Roll dice → Place batch into process

N → Wait

The Lean Games and Simulations Book: Second Edition. Copyright © John Bicheno, 2014

# Single Piece Flow Dice Game
## (5 Similar Workstations Example)

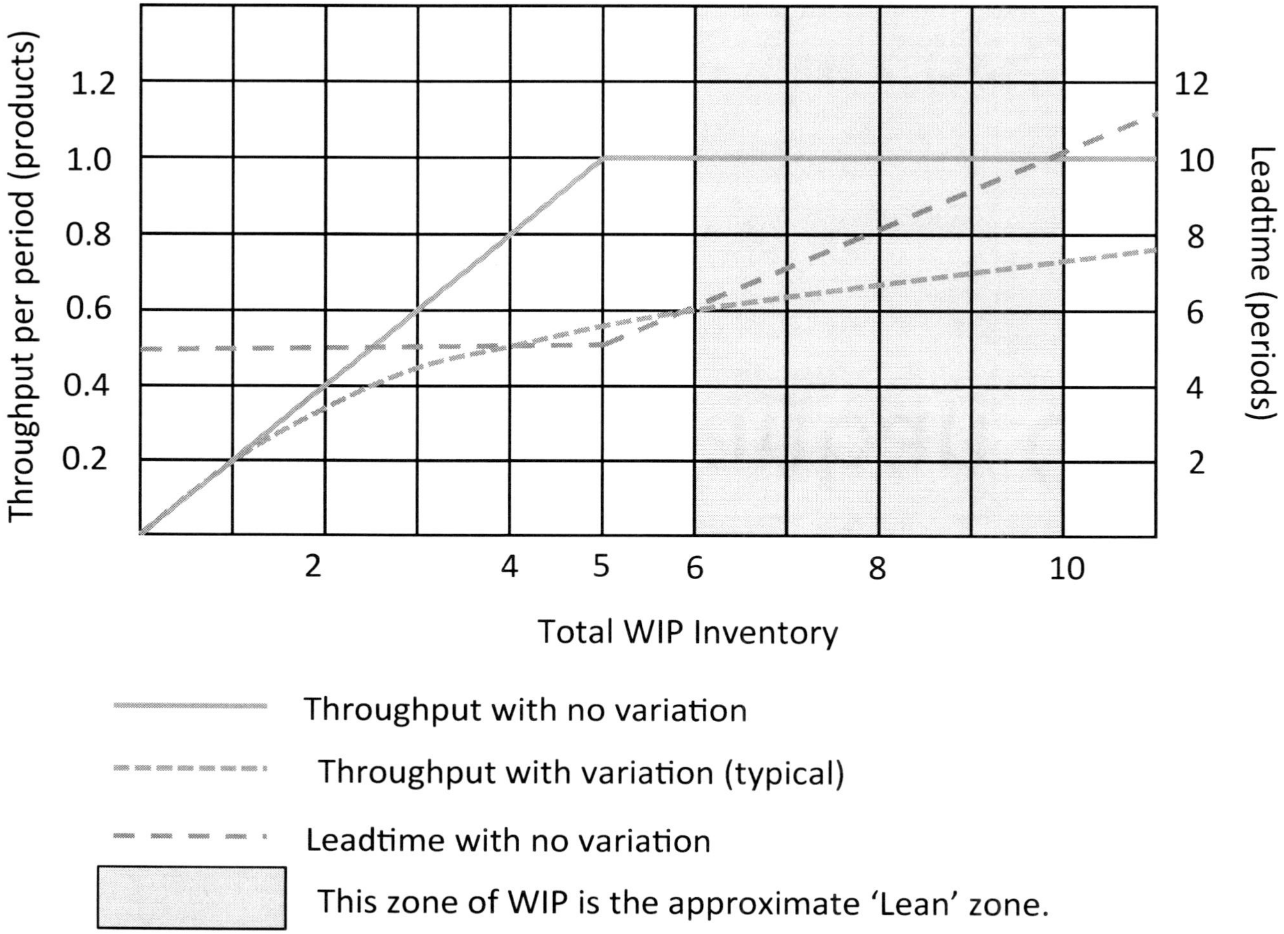

Throughput per period (products)
1.2
1.0
0.8
0.6
0.4
0.2
12
10
8
6
4
2
Leadtime (periods)
2
4
5
6
8
10
Total WIP Inventory
Throughput with no variation
Throughput with variation (typical)
Leadtime with no variation
This zone of WIP is the approximate ‘Lean’ zone.


- 5 Stations, each with an average cycle time of one full period (2 half periods)
- No bottleneck
- Bottleneck rate (BNR) = 1
- Critical WIP = BNR * (sum of process times) = 1 * 5
- At critical WIP with no variation, throughput is at a maximum and leadtime is at a minimum.
- With variation, throughput drops and leadtime increases.
- Throughput with variation line shows “Practical Worst Case’ (Factory Physics)

The Lean Games and Simulations Book: Second Edition. Copyright © John Bicheno, 2014

# Kanban Games

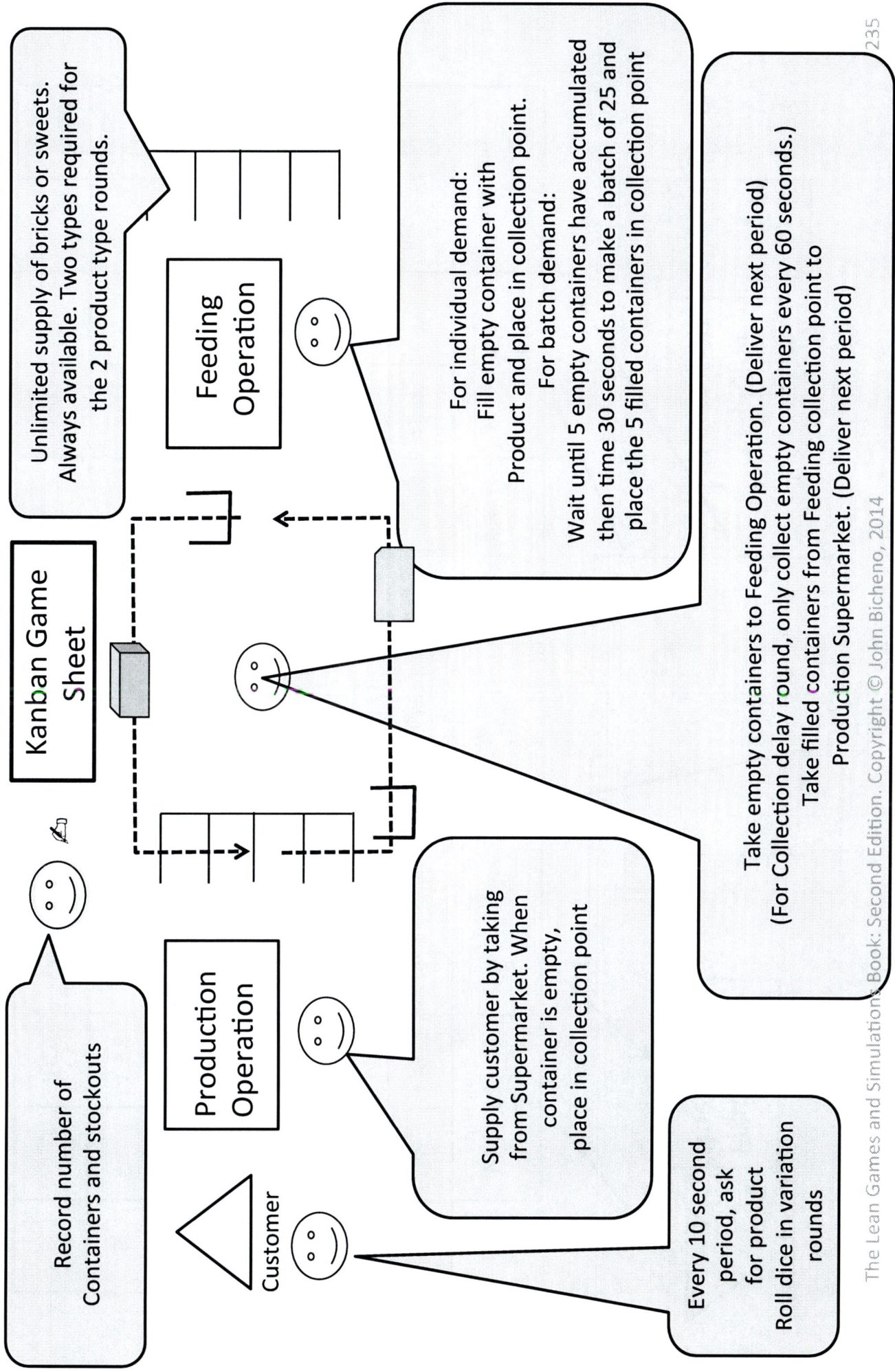

Unlimited supply of bricks or sweets. Always available. Two types required for the 2 product type rounds.

Kanban Game Sheet

Feeding Operation

For individual demand:
Fill empty container with Product and place in collection point.
For batch demand:
Wait until 5 empty containers have accumulated then time 30 seconds to make a batch of 25 and place the 5 filled containers in collection point

Take empty containers to Feeding Operation. (Deliver next period)
(For Collection delay round, only collect empty containers every 60 seconds.)
Take filled containers from Feeding collection point to Production Supermarket. (Deliver next period)

Record number of Containers and stockouts

Production Operation

Supply customer by taking from Supermarket. When container is empty, place in collection point

Customer

Every 10 second period, ask for product
Roll dice in variation rounds

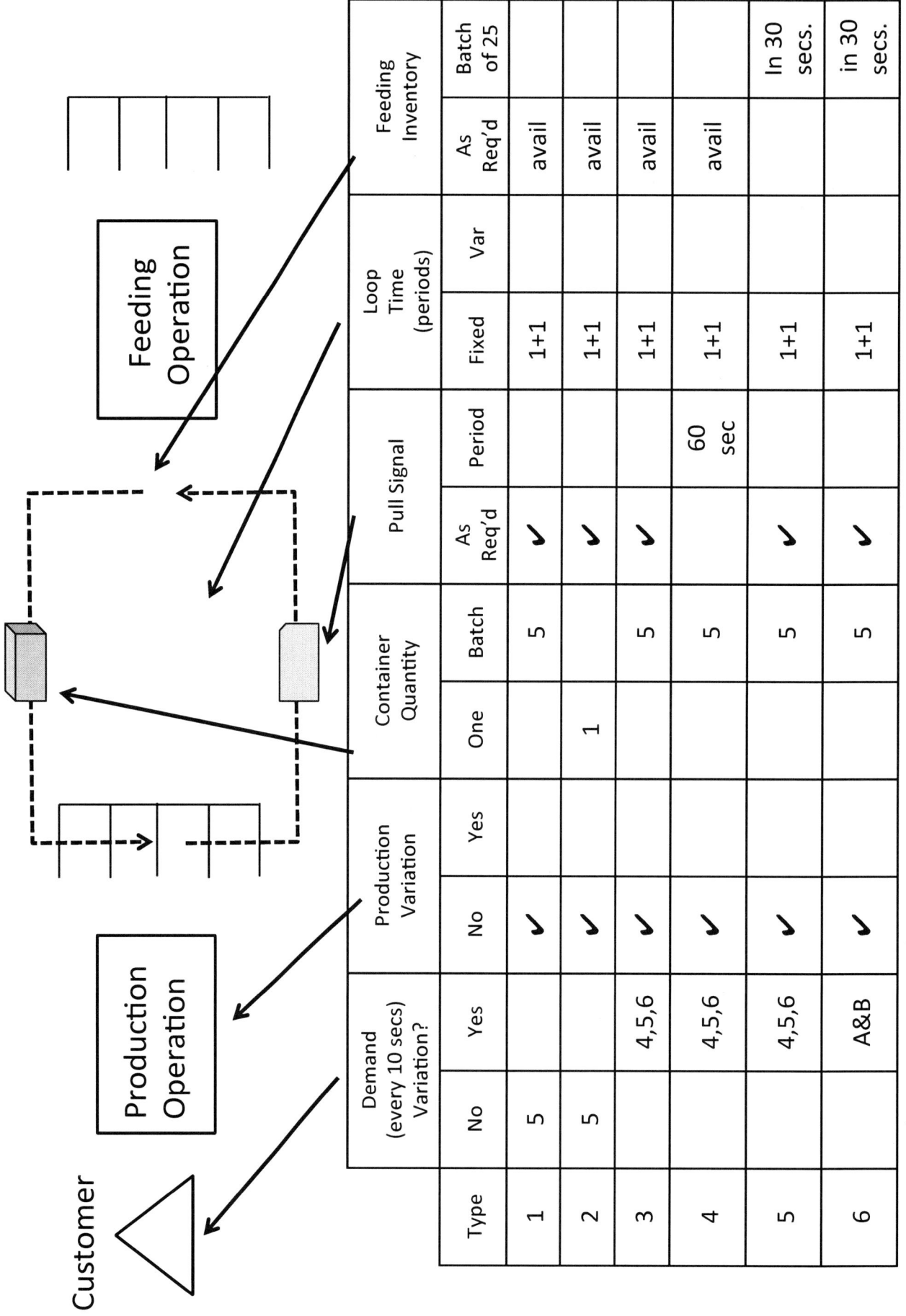

| Type | Demand (every 10 secs) Variation? | | Production Variation | | Container Quantity | | Pull Signal | | Loop Time (periods) | | Feeding Inventory | |
|---|---|---|---|---|---|---|---|---|---|---|---|---|
| | No | Yes | No | Yes | One | Batch | As Req'd | Period | Fixed | Var | As Req'd | Batch of 25 |
| 1 | 5 | | ✔ | | | 5 | ✔ | | 1+1 | | avail | |
| 2 | 5 | | ✔ | | 1 | | ✔ | | 1+1 | | avail | |
| 3 | | 4,5,6 | ✔ | | | 5 | ✔ | | 1+1 | | avail | |
| 4 | | 4,5,6 | ✔ | | | 5 | | 60 sec | 1+1 | | avail | |
| 5 | | 4,5,6 | ✔ | | | 5 | ✔ | | 1+1 | | | In 30 secs. |
| 6 | | A&B | ✔ | | | 5 | ✔ | | 1+1 | | | in 30 secs. |

Number of cards =

$$\frac{(\text{Demand} \ \times \ \text{Lead Time}) \ + \ \text{Risk Quantity}}{\text{Container Quantity}}$$

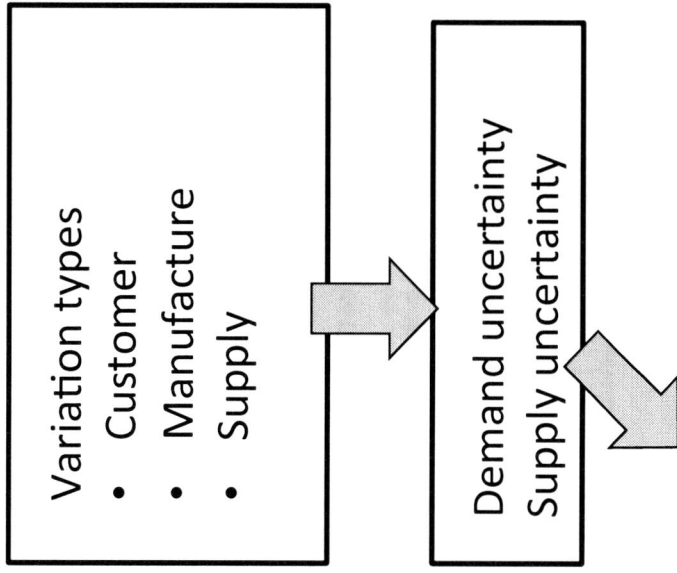

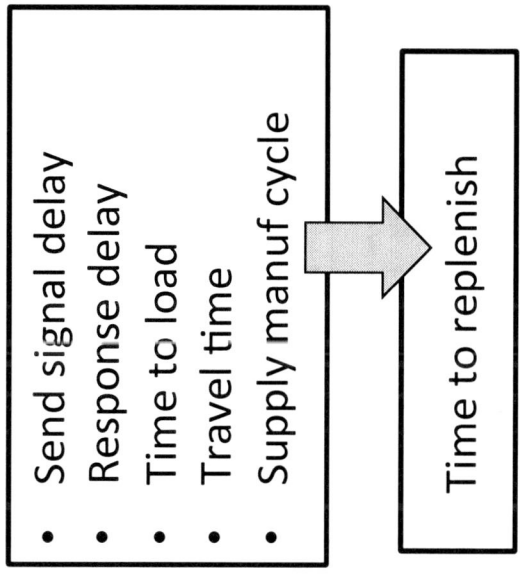

Variation types
- Customer
- Manufacture
- Supply

➡ Demand uncertainty
Supply uncertainty

- Send signal delay
- Response delay
- Time to load
- Travel time
- Supply manuf cycle

➡ Time to replenish ➡

Average Demand During Lead Time

Number of cards =

$$\frac{(\text{Demand} \ \times \ \text{Lead Time}) \ + \ \text{Risk Quantity}}{\text{Container Quantity}}$$

# Kanban Games

| Type | Demand (every 10 secs) Variation? — No | Demand — Yes | Production Variation — No | Production — Yes | Container Quantity — One | Container — Batch | Pull Signal — As Req'd | Pull Signal — Period | Loop Time (periods) — Fixed | Loop Time — Var | Feeding Inventory — As Req'd | Feeding Inventory — Batch of 25 |
|------|------|------|------|------|------|------|------|------|------|------|------|------|
| 1 | 5 | | ✔ | | | 5 | | | 1+1 | | avail | |
| 2 | 5 | | ✔ | | 1 | | ✔ | | 1+1 | | avail | |
| 3 | | 4,5,6 | ✔ | | | 5 | ✔ | | 1+1 | | avail | |
| 4 | | 4,5,6 | ✔ | | | 5 | | 60 sec | 1+1 | | avail | |
| 5 | | 4,5,6 | ✔ | | | 5 | ✔ | | 1+1 | | | In 30 secs. |
| 6 | | A&B | ✔ | | | 5 | ✔ | | 1+1 | | | in 30 secs. |

# Notes on Kanban Game

- 6 games of 5 minutes each with decisions before and review after every game
- Several parallel games
- Instructor can demonstrate game 1
- There is a competition for lowest number of containers in the loop, but disqualified if there is a supply failure to the customer
- Teams are given the basic formula, but must decide on the number of kanbans (or containers) and initial containers at the Production supermarket.
- The Kanban 'cards' are the containers.
- Kanban quantity is fixed at 5 products
- Use wrapped sweets or 4 stud lego for inventory; paper cups or plastic boxes for containers; Two types or colour required for two product round. Two timers required per team.
- 5 Players per team: Customer, Production, Material Handler, Supplier, Scribe
- Each game has 5min x 6 pulls = 30 pulls of 10 seconds
- Kanban loop has two legs, Collection and Delivery, each of 10 seconds.
- Demand variation generated by dice: 4, 5, 6 are valid; 1=4, 2=5, 3=6
- Round 4 simulates a runner or tugger system, collecting kanbans every 60 seconds.
- Round 5 simulates internal supply where a previous workstation takes 30 seconds to make a batch of 25 products.
- Round 6 has two products.

# A Useful Symbol Reminder

The (slightly modified) Supermarket Icon
Is a useful reminder:

- The customer pull quantity

- The make batch size

- Buffer stock (for manufacture variation)

- Safety stock (for customer variation)

Add these together to calculate the inventory

In the Loop

Inventory in The Loop can be in several places:
In the pre-work supermarket, in a container en route, as a kanban en route or
Being accumulated. The latter two are virtual inventory.

# CONWIP and Drum Buffer Rope Explorer Exercise

- CONWIP and DBR are multi-stage pull systems, often far more effective than stage-by-stage kanban.

- In CONWIP and DBR a quantity is released at the start of the process equal to the quantity that has just been completed - at the last process (CONWIP) or at the bottleneck (DBR).

- The (earlier) dice games give practice in CONWIP and DBR

- But in the real world it is not so simple:
  - When is the signal to release given? (immediately, daily, weekly?
  - What if there are batch quantities that need to be made together?
  - What if there is more than one product?
  - What happens if different orders have different priorities?

- This exercise explores various possibilities, leading to a greater appreciation of the options available.

The Lean Games and Simulations Book: Second Edition. Copyright © John Bicheno, 2014

# CONWIP and DBR Explorer Order Cards

| Product A DD = 36 | Product A DD = 36 | Product A DD = 38 | Product A DD = 38 | Product A DD = 40 | Product A DD = 40 |
|---|---|---|---|---|---|
| Product A DD = 42 | Product A DD = 42 | Product A DD = 44 | Product A DD = 44 | Product A DD = 46 | Product A DD = 46 |
| Product A DD = 48 | Product A DD = 48 | Product A DD = 50 | Product A DD = 50 | Product A DD = 52 | Product A DD = 52 |
| Product B DD = 36 | Product B DD = 36 | Product B DD = 38 | Product B DD = 38 | Product B DD = 40 | Product B DD = 40 |
| Product B DD = 42 | Product B DD = 42 | Product B DD = 44 | Product B DD = 44 | Product B DD = 46 | Product B DD = 46 |
| Product B DD = 48 | Product B DD = 48 | Product B DD = 50 | Product B DD = 50 | Product B DD = 52 | Product B DD = 52 |

# CONWIP and DBR Explorer Game Sheet

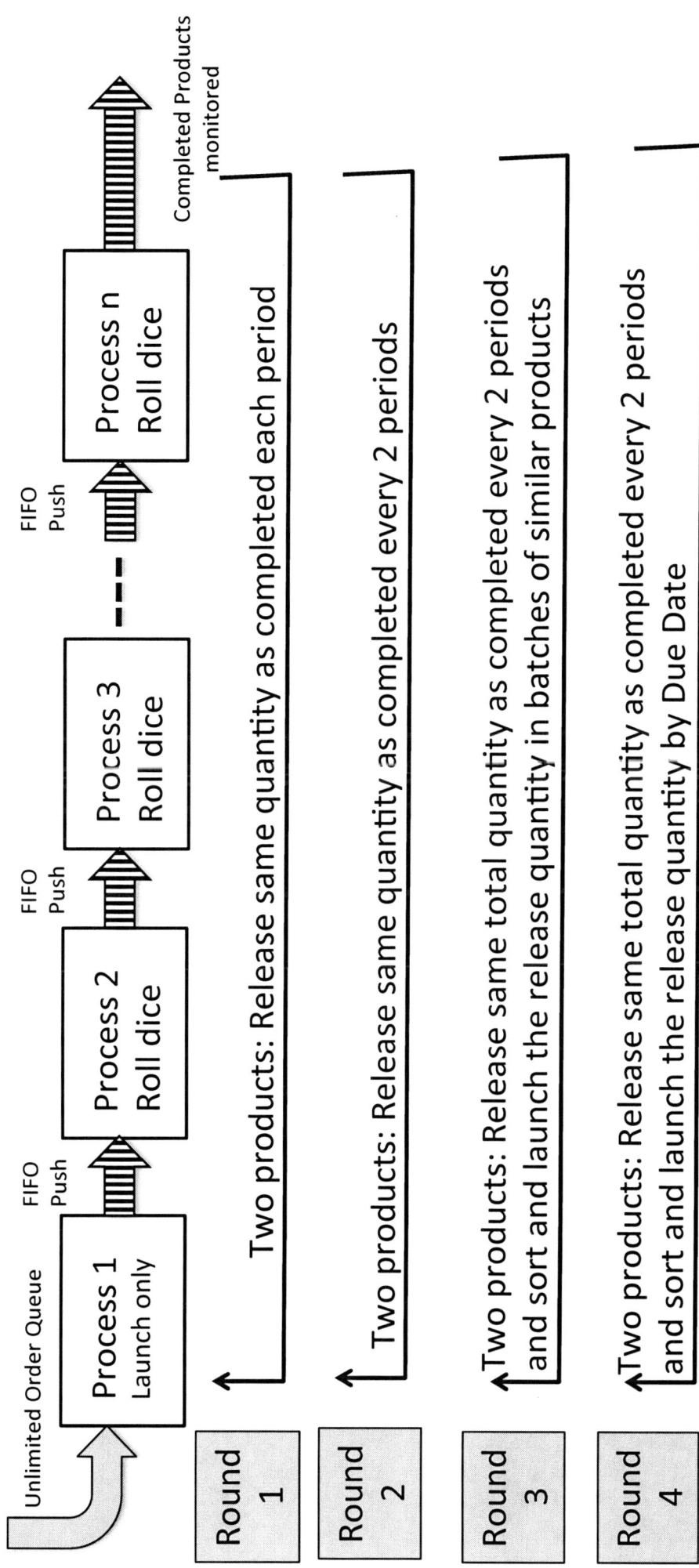

Unlimited Order Queue

**Process 1**
Launch only

FIFO
Push

**Process 2**
Roll dice

FIFO
Push

**Process 3**
Roll dice

FIFO
Push

**Process n**
Roll dice

Completed Products monitored

**Round 1** — Two products: Release same quantity as completed each period

**Round 2** — Two products: Release same quantity as completed every 2 periods

**Round 3** — Two products: Release same total quantity as completed every 2 periods and sort and launch the release quantity in batches of similar products

**Round 4** — Two products: Release same total quantity as completed every 2 periods and sort and launch the release quantity by Due Date

Product sequence remains the same for all rounds.
Initial WIP at beginning of each round is 4 products at each process

# CONWIP and DBR Explorers
## Some lessons?

1. CONWIP and DBR limit inventory

2. It is often not practical to release as completed, so a release period must be decided on. Inventory is limited to an upper bound, but WIP may need to increase to reduce risk of starvation upstream. The assumption that all jobs are equally important is often not true.

3. As for 2. Batching has due date consequences. Makes due date performance even worse. But a trade-off particularly if there is changeover. Lean mixed model is much better, so strive for this. Sorting jobs makes order promising more difficult, and scheduling more challenging.

4. As for 2. Due Date sorting improves delivery performance. Scheduling is more complex.

5. Other extensions are possible, but all have 'good and bad news': For instance
   - Sorting by due date and batches (This combines 3 and 4)
   - Sorting by shortest processing time (Some jobs are early, others very late)
   - Sorting by slack rule (often done with MRP: the difference in time between when a job is due and the time that it will take to complete it. This may be more applicable in job shop situations, but scheduling is more complex.)

# Cut and Fill Games

# Cut and Fill Game

- Initial Round (Push System)
- Line Balancing
- Kanban
- CONWIP
- Self Balancing
- Discussion

The Lean Games and Simulations Book: Second Edition. Copyright © John Bicheno, 2014

# Cut and Fill: Flow Diagram

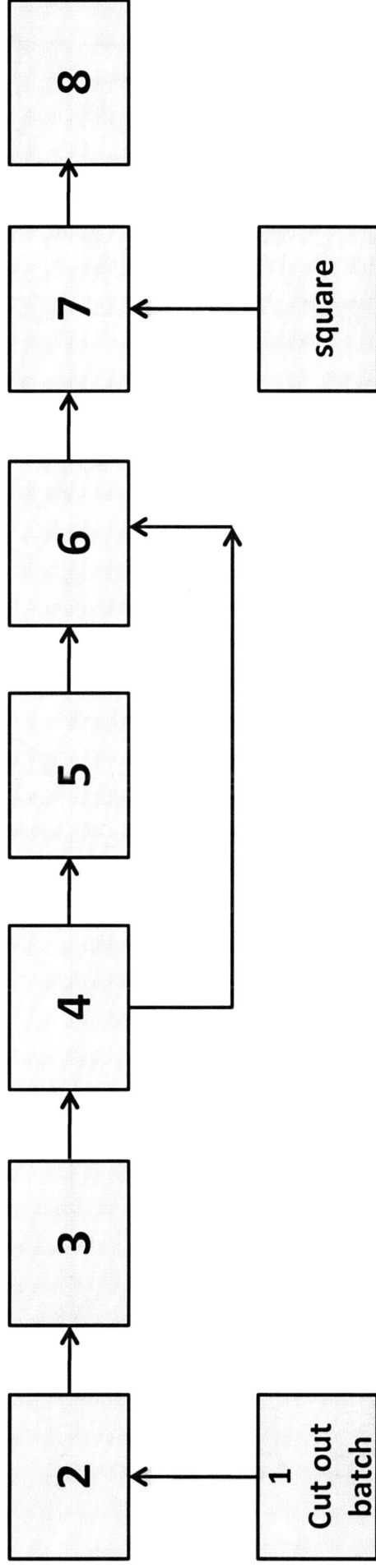

| 1 Cut out batch | → | 2 | → | 3 | → | 4 | → | 5 | → | 6 | → | 7 | → | 8 |

square → 7

4 → 6

Initial player Allocation: One per Station,
plus a Material Handler, and an Industrial Engineer

# Cut and Fill: Player Card (for each player)

Product Number

| 3 | 4 | 5 | 6 | 7 | 8 |
|---|---|---|---|---|---|

1. Read schedule. Cut out the indicated quantity of blank product squares from the Product Card strips. Do not cut out individual squares.

2. Copy the outline shape onto the product card

3. Colour Red or Blue
   ** 10 second change between colours **

4. Colour Green (not done for products 4, 5, 7)

5. PRINT number inside

6. Cut out shapes

7. Glue the shape onto a product square

8. Inspect

The Lean Games and Simulations Book: Second Edition. Copyright © John Bicheno, 2014

# Cut and Fill Game: Player Instructions

- Station 1: Read the schedule. Cut out the required quantity (batch) of products from the Product Squares sheet. Do not cut out individual products.

- Station 2: Draw the shapes onto each square.

- Station 3: Colour in Red or Blue as shown. If (when) you change colours you must time 10 seconds.

- Station 4: Colour in Green as shown.

- Station 5: Write (PRINT) the appropriate product number inside each square. Use words not numbers.

- Station 6: Cut out the shapes.

- Station 7: Stick the shape onto a blank square.

- Station 8: Inspect: See that colours are correct and that shapes are properly coloured in. Use your judgment.

# Cut and Fill Game: Materials

- Station 1: Scissors.
- Station 2: Black felt pen
- Station 3: Black felt pen
- Station 4: Red and Blue felt pens. Timer
- Station 5: Green felt pen
- Station 6: Scissors
- Station 7: Glue Stick and Scissors
- Station 8: nil
- All stations need a copy of the Player Card.
- Note: Select fairly thick felt pens; thin pens slow down the rate of work.

# Cut and Fill Product Squares

(instructor to cut into strips of four)

The Lean Games and Simulations Book: Second Edition. Copyright © John Bicheno, 2014

# Cut and Fill Game: Customer Order Sequence

| Order No | Product no | Batch Quantity | | Order No | Product No | Batch Quantity |
|---|---|---|---|---|---|---|
| 1 | 4 | 4 | | 13 | 5 | 2 |
| 2 | 3 | 2 | | 14 | 4 | 5 |
| 3 | 7 | 3 | | 15 | 6 | 3 |
| 4 | 5 | 1 | | 16 | 7 | 1 |
| 5 | 6 | 4 | | 17 | 3 | 2 |
| 6 | 4 | 3 | | 18 | 4 | 3 |
| 7 | 7 | 2 | | 19 | 7 | 3 |
| 8 | 8 | 1 | | 20 | 6 | 2 |
| 9 | 3 | 4 | | 21 | 5 | 1 |
| 10 | 8 | 3 | | 22 | 3 | 3 |
| 11 | 5 | 2 | | 23 | 7 | 2 |
| 12 | 4 | 3 | | 24 | 4 | 2 |

# Cut and Fill: Balance Table

| Operation | Activity Time | Adjusted Time | Grouping |
|---|---|---|---|
| 1 | | | |
| 2 | | | |
| 3 | | | |
| 4 | | | |
| 5 | | | |
| 6 | | | |
| 7 | | | |
| 8 | | | |

# Operator Balance Board

| Time (seconds) | Op 1 | Op 2 | Op 3 | Op 4 | Op 5 | Op 6 | Op 7 | Op 8 |
|---|---|---|---|---|---|---|---|---|
| 20 | | | | | | | | |
| 18 | | | | | | | | |
| 16 | | | | | | | | |
| 14 | | | | | | | | |
| 12 | | | | | | | | |
| 10 | | | | | | | | |
| 8 | | | | | | | | |
| 6 | | | | | | | | |
| 4 | | | | | | | | |
| 2 | | | | | | | | |
| 0 | | | | | | | | |

# Kanban Square

One of these is required between each pair of Operators

The Square may contain only one Product.
When the square is occupied the previous operation must stop work.
When the square is empty the previous operation is authorised to begin work to fill the square.

# FIFO Lane

One of these is required between Complex Stages and where there is Changeover

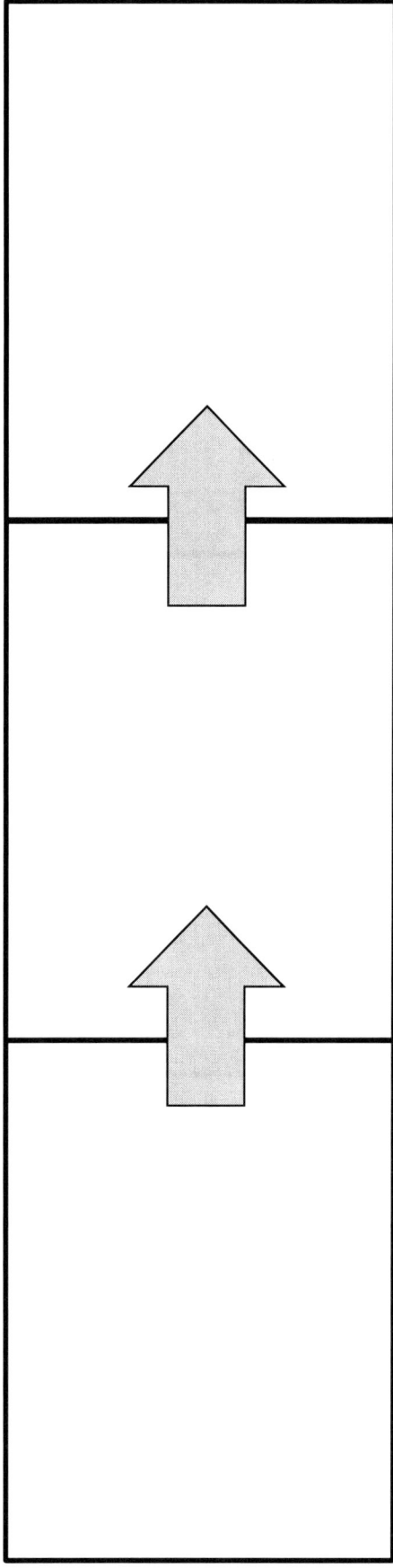

The FIFO lane may contain only three Products.

Products must move up the line from left to right

When the leftmost square is occupied the previous operation must stop work.

When the leftmost square is empty the previous operation is authorised to begin work to fill the square.

# Cut and Fill Game: Measures

| Round | Completed | WIP | Rejects | Lead time | Players | Productivity |
|---|---|---|---|---|---|---|
| Basic | | | | | | |
| Balanced Line | | | | | | |
| Kanban | | | | | | |
| CONWIP | | | | | | |
| Bucket Brigade | | | | | | |

# Cut and Fill: Balanced (?)

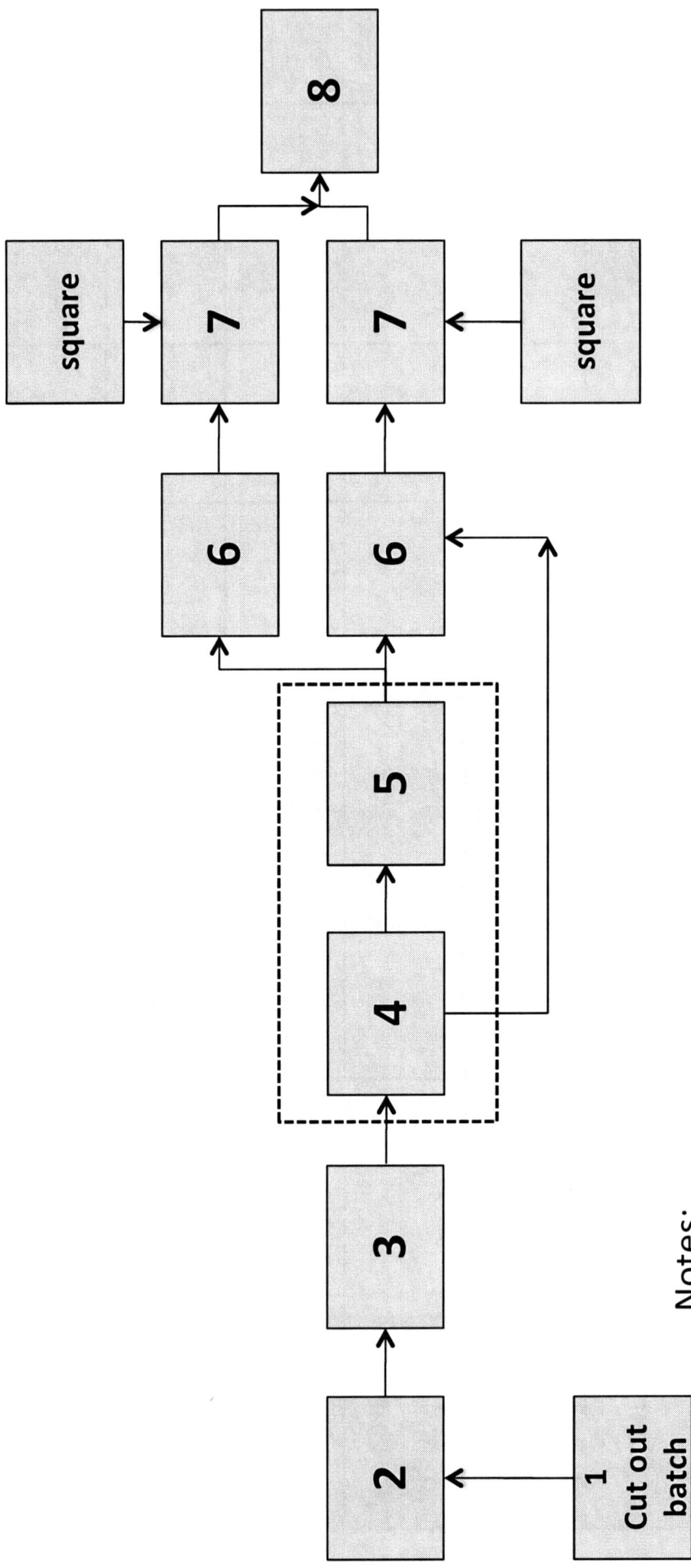

Notes:

4 and 5 done by 1 operator; 6 and 7 each done by 2 operators

Can operators 7 also do inspection?

# 5S Numbers Game

45

93

30

62

1    65

49

81

33

17

18

82

51

34

53    85

21

69    37

98

54

70

38

89

25

73    9

26

74

58

90

10

57

41

5

6

22

39    87

23    55    91

7    71    3

50

11

67    19

83

35    36

52    4

84

47

95    15

31

63    75    79

27    59

43

88

68    70

56    8    40    24    72

32

94    46

78

64

16    80

48

28    76

92    44

60    12

20

96

45

30

62

15

32

1

29

61

13

47

46

14

63

31

64

48

9

25

42

58

27

59

44

28

5

41

57

26

10

43

3

60

12

53

37

21

6

22

54

23

55

39

7

40

8

24

38

19

50

11

56

20

33

49

51

36

35

52

4

34

2

18

17

49  17  33  1
53  21  5  37  41
57  9  25

18  2  34
54  22  6  38
26  10  42  58

50  11  19
7  23  55  39
27  3  43  59

52  4  36
56  24  8  40
28  12  60  44

45  29  61  13
30  62  14  46
47  63  31  15  16
32  64  48

| 1 | 2 | 3 | 4 | 5 | 6 | 7 | 8 |
|---|---|---|---|---|---|---|---|
| 9 | 10 | 11 | 12 | 13 | 14 | 15 | 16 |
| 17 | 18 | 19 | 20 | 21 | 22 | 23 | 24 |
| 25 | 26 | 27 | 28 | 29 | 30 | 31 | 32 |
| 33 | 34 | 35 | 36 |  | 38 | 39 | 40 |
| 41 | 42 | 43 | 44 | 45 | 46 | 47 | 48 |
| 49 | 50 | 51 | 52 | 53 | 54 | 55 | 56 |
| 57 | 58 | 59 | 60 | 61 | 62 | 63 | 64 |

Sheet 4

45 93 30 62 32 96
29 61 94 15
77 46 95 16
13 95 79 80
14 31 48
78 63 64

89 42 76
73 25 58 28
9 74 27 75 44
41 90 59 92
57 26 10 3
43 91 60
91 12

85 5 6 22 24
17 21 37 98 23 72
50 54 71 8
33 51 7 56
82 38 11 55 68
2 66 39 2
18 87 36
65 83 40 52
49 67 88 4
81 19 84
34 35 84

Sheet 1: Instructor only

Sheet 2: Instructor only

# Notes on 5S Numbers Sheets

- No missing numbers, except in final slide
- 50 and 51 interchanged
- 3 and 11 interchanged
- Font of 19 smaller
- Font of 6 bigger

# Advantages of 5S

- Productivity (Speed)
- Reduced variation
- Problem identification
- Improved morale very likely

# The Manufacturing Cell

# Manufacturing Cell: Initial Layout

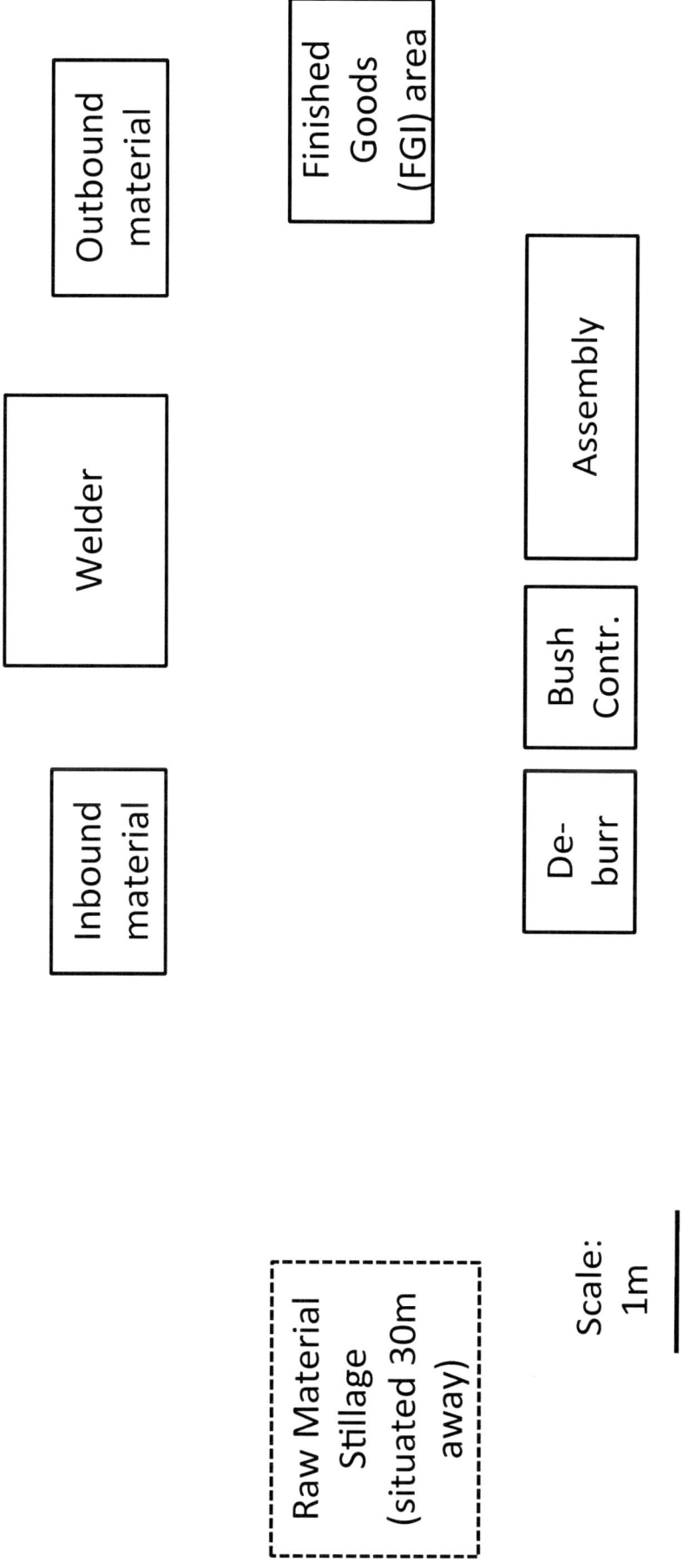

Inbound material

Welder

Outbound material

Finished Goods (FGI) area

De-burr

Bush Contr.

Assembly

Raw Material Stillage (situated 30m away)

Scale: 1m

# Manufacturing Cell Activities

The demand from a process is 800 products per day. The area works an 8 hour day, with a lunch break of 30 minutes, two 20 minute refreshment breaks, and an additional 10 minutes allowed for comfort.

| Activity | Description | Time |
|---|---|---|
| A | Fetch inbound material from raw material stillage (every 30th cycle) | 120 |
| B | Walk to inbound raw material container | 5 |
| C | Pick up raw material | 2 |
| D | Walk to left side of weld machine | 3 |
| E | Place raw material on left of weld machine | 2 |
| F | Walk to centre of weld machine | 2 |
| G | Remove welded part from jig on weld machine | 5 |
| H | Place welded part in outbound area | 2 |
| I | Walk to left side of weld machine | 3 |
| J | Pick up raw material from left side of weld machine | 3 |
| K | Move raw material to centre of weld machine | 2 |
| L | Secure raw material into jig | 6 |
| M | Start welder | 2 |
| N | Welder runs through cycle | 18 |
| P | Wait for part to cool | 15 |
| Q | Pick up welded part and jig from weld machine | 2 |
| R | Visual inspection of part | 6 |
| S | Carry part to bush area (3 out of 4 parts on average) | 3 |
| T | Carry part to de-burr area (every 4th part on average) | 4 |
| U | De-burr (every 4th part on average) | 12 |
| V | Carry de-burred part to bush area (every 4th part) | 2 |
| W | Fetch bushes container (every 200 cycles approx) | 200 |
| X | Pick up bush | 2 |
| Y | Walk to assembly table | 3 |
| Z | Assemble product | 10 |
| A1 | Place assembled product in finished goods (FGI) area | 3 |
|   |   |   |

# Cell Game
## Balance Board

Each line is one second

Draw in takt time and planned cycle time lines

| Time | Op 1 | Op 2 | Op 3 | Op 4 | Op 5 | Op 6 |
|------|------|------|------|------|------|------|
| 0 | | | | | | |
| 2 | | | | | | |
| 4 | | | | | | |
| 6 | | | | | | |
| 8 | | | | | | |
| 10 | | | | | | |
| 12 | | | | | | |
| 14 | | | | | | |
| 16 | | | | | | |
| 18 | | | | | | |
| 20 | | | | | | |
| 22 | | | | | | |
| 24 | | | | | | |
| 26 | | | | | | |
| 28 | | | | | | |
| 30 | | | | | | |
| 32 | | | | | | |
| 34 | | | | | | |

# The Paperwork (Office) Cell

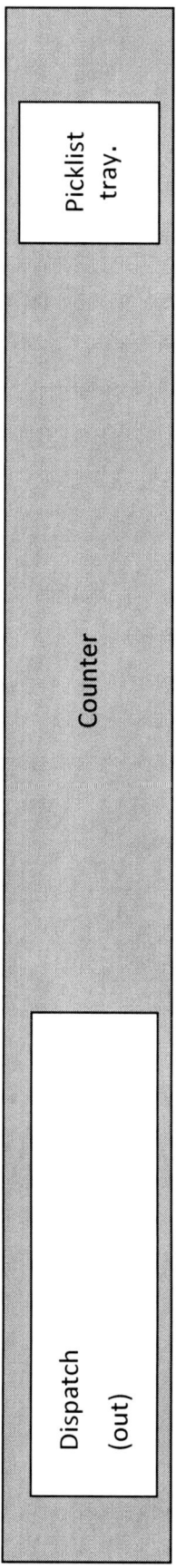

Picklist tray.

Counter

Dispatch
(out)

The
Office

Filing
cabinet

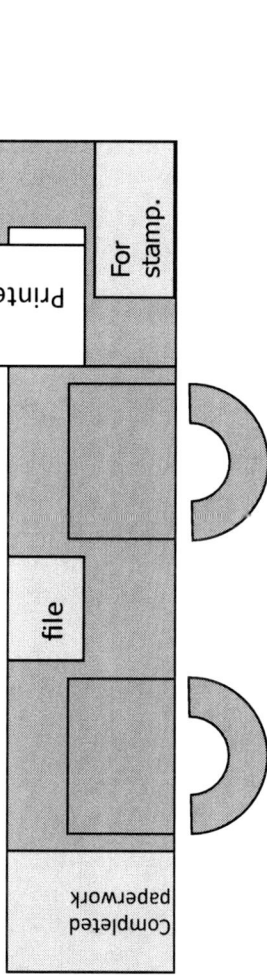

In

Computer

Printers

For
stamp.

file

Completed
paperwork

1 metre

Note: The Warehouse is 'North' of this area

# Basic Cell Analysis Tools

- Takt time
- One piece flow
- 'Value Add' and Non Value Add activities
- Work Balance Board
- Spaghetti Diagram
- Standard Work Combination table
- Standard Operating Procedure

# The Paperwork Cell

- The Paperwork Cell is the last stage before final dispatch in a stationery company that takes orders over the phone and internet, picks the orders, and ships overnight. Orders are received throughout the day. Average demand is around 760 orders per day, but is expected to rise by around 100 per day. The office works an 8 hour day with one 30 minute break and two 10-minute breaks. The following is a process activity map collected from video sequences. The times are median times in seconds for processing a single order. Assume that the workload is evenly distributed throughout the day. (This is unrealistic, so you should build in an allowance for variation).

- What you need to do:
  - Calculate takt time (based on anticipated figures)
  - Calculate the required number of office workers for one piece flow
  - Go through the list of activities and identify the 'value adding' steps
  - Decide on an appropriate target number of office workers
  - Examine the activities and layout critically, and make improvements. Show the new layout on a Layout Diagram
  - Show the revised way of working on a Work Balance Board

- What is allowed and what is not
  - Anything can be changed or rearranged, but
  - No equipment can be placed on the counter for security and safety reasons
  - A paper invoice is still required

# Cell Activity List

| Activity No | Unit Time (secs) | Activity |
|---|---|---|
| A | 3 | Walk to picklist box on counter |
| B | 2 | Pick up picklist |
| C | 3 | Reorient and sort picklist |
| D | 3 | Walk to desk |
| E | 1 | Place picklist in in-basket |
| F | 3 | Walk to computer chair and sit down |
| G | 2 | Pick up picklist from in-basket |
| H | 8 | Scan picklist bar code and wait for order to be retrieved on computer screen |
| I | 15 | Check picklist against original order. All items picked ? (Y/N : if Y (80%) go to activity L |
| J | 24 | N: Enter changes to invoice on computer (20%). |
| K | 8 | Enter reason code  for shortage into computer |
| L | 3 | Check paper, insert envelope and start printer |
| M | 15 | Print out 2 copies of the invoice, and an envelope |
| N | 3 | Separate office and customer copies of the invoice |
| O | 6 | Stamp 'checked' , date and clerk number  on customer copy |
| P | 4 | Place customer copy of invoice into envelope |
| Q | 2 | Place envelope into customer invoice tray |
| R | 2 | Place office copy into file tray |
| S | 8 | (Every approx 10 envelopes)  Walk from customer tray to dispatch box, walk back |
| T | 15 | (Every approx 10 copies) Take office copy to filing cabinet and file, walk back |

# Takt Time

- The average rate that work should come out of a cell or line

- Available time / demand

- In an office this gives an idea of the rate of work

- Useful for work allocation and for calculating the number of workers

- In an office, takt is a guideline. No-one expects that workers to complete an order every takt cycle, but over a few hours the takt will approximate the rate

- In an office, you should make allowance for demand variation and work variation - perhaps balance to 80% of takt (90%+ in a factory).

- This increased time (takt /.80) is known as planned cycle time.

# 'Value adding' and 'waste'

- Strictly, a value adding activity is something that a customer is prepared to pay for. If it does not add value it is 'waste'

- In an office, 'value add' is something that managers or customers would be prepared to pay a third-party provider to do, were this service available.

- In some offices and service situations it may be good enough to define waste as anything that prevents a job being done immediately and right first time.

- All 'waste' activities should be eliminated or reduced

- Put all waste activities under the microscope. Question them!

# One-Piece Flow

- Moving pieces between stages one at a time rather than in batches

- Batches may appear to be efficient, but penalties are paid in
  - Completion Time (Lead time)
  - Space
  - Quality (where a defect is noticed, and rework is required)

- One piece flow often involves more walk and setup time, so these must be attacked.

# Work Balance Board

- A board, displayed at the workplace that helps allocate work elements between operators
- Mark in the takt time
- Mark in the target balance cycle (planned cycle time)
- Accumulate times for each operator
- Colour waste activities differently

Op1 Op2

| Time | Op 1 | Op 2 | Op 3 | Op 4 | Op 5 | Op 6 |
|------|------|------|------|------|------|------|
| 0    |      |      |      |      |      |      |
| 2    |      |      |      |      |      |      |
| 4    |      |      |      |      |      |      |
| 6    |      |      |      |      |      |      |
| 8    |      |      |      |      |      |      |
| 10   |      |      |      |      |      |      |
| 12   |      |      |      |      |      |      |
| 14   |      |      |      |      |      |      |
| 16   |      |      |      |      |      |      |
| 18   |      |      |      |      |      |      |
| 20   |      |      |      |      |      |      |
| 22   |      |      |      |      |      |      |
| 24   |      |      |      |      |      |      |
| 26   |      |      |      |      |      |      |
| 28   |      |      |      |      |      |      |
| 30   |      |      |      |      |      |      |
| 32   |      |      |      |      |      |      |
| 34   |      |      |      |      |      |      |

Spaghetti

Picklist tray.

Counter

In

Computer

Printers

For sign.

file

Completed paperwork

Dispatch (out)

Filing cabinet

1 metre

Note: The Movements above are for illustration only

# Service Call Centre Operations

# Aims

- Demonstrate Failure Demand and advantages of reducing Failure Demand

- Get a feel for the statistics of variation in a service centre, specifically

  - The problem of averages, and average-based KPI's
  - The skewed Poisson distribution
  - The need for safety utilization factors (below 100%)
  - Non-linear queuing effects

## Call Centre Game Flow Sheet

**Work Complete**

**Call abandoned by operator**

**Multiple Parallel operators**

**Caller Tries again**

**Queuing**

**Calls Arrive Every 15 seconds (Poisson)**

Place abandoned slips on bottom of queue

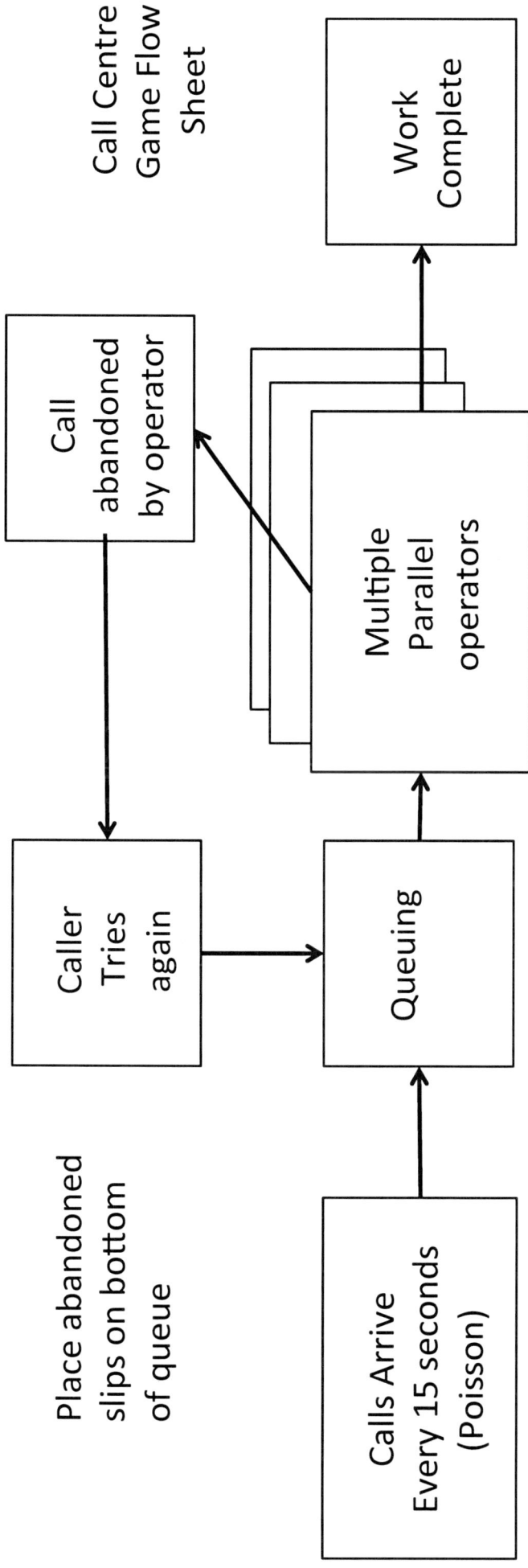

If not dealing with a current case, take next slip.
Each operator rolls a dice twice per period.
(For a new or repeat customer, write current period and do one dice roll during this period)
Keep tags on rolls per customer (use gate IIII ⫫)
If a 6 is rolled, the case is completed (write current period on slip, pass to complete).
If 1 to 5 is rolled, more rolls are required to complete the call.
After X rolls the case is abandoned (write 'A', and place the slip at the bottom of the queue.

Play for 40 periods
Each period, look up number of arrivals
On slip of paper, write
- Customer number
- Arrival period
Place slips on bottom of queue (a FIFO lane)

The Lean Games and Simulations Book: Second Edition. Copyright © John Bicheno, 2014

# Notes on Service Operations / Call Centre Game

- Best to have at least two parallel games

- Play each game at least twice – with different levels of abandonment. That is, the case is 'abandoned' after X number of dice rolls. This would simulate a call centre or operations target where long calls are discouraged.

- Use a PowerPoint display to show the current period. Alternatively prepare a flipchart with 40 periods and have someone indicate the current period. (Each period is 15 seconds).

- Prepare 80 (or 120 slips for the larger game) paper slips per round per game – so minimum 160.

- Players are told that their game will have 2 arrivals per period on average. (The instructor may decide on 1 or 3 arrivals per period if there are insufficient or sufficient players.)

- A guideline: around 8 or more operators for two calls per period; around 11 or more for three calls per period, but an additional two people are required for each game for game facilitation.

- Operators work in parallel

- There is one queue – a FIFO lane

- The teams should decide how many operators they require. (The teams should have different numbers of operators)

- Teams should also decide on the limit number of dice rolls before a call is abandoned.

- Measures: Average number of periods to complete end-to-end; number of abandoned calls (failure demands); Longest end-to-end time (in periods),

- Calculate utilization: = load / capacity = ave calls per period / (no of operators x rolls per period / ave rolls for a 6). So for 1 call per period, 3 operators, 6 rolls for a 6, utilization = 100%.

# Poisson Distribution for 1 arrival per period: 5 Games each of 40 periods.

| period | no of arrivals | cum arrivals |
|---|---|---|
| 1 | 1 | 1 |
| 2 | 1 | 2 |
| 3 | 2 | 4 |
| 4 | 1 | 5 |
| 5 | 0 | 5 |
| 6 | 2 | 7 |
| 7 | 1 | 8 |
| 8 | 0 | 8 |
| 9 | 1 | 9 |
| 10 | 1 | 10 |
| 11 | 1 | 11 |
| 12 | 2 | 13 |
| 13 | 2 | 15 |
| 14 | 1 | 16 |
| 15 | 0 | 16 |
| 16 | 4 | 20 |
| 17 | 0 | 20 |
| 18 | 1 | 21 |
| 19 | 1 | 22 |
| 20 | 0 | 22 |
| 21 | 1 | 23 |
| 22 | 3 | 26 |
| 23 | 1 | 27 |
| 24 | 1 | 28 |
| 25 | 1 | 29 |
| 26 | 2 | 31 |
| 27 | 1 | 32 |
| 28 | 1 | 33 |
| 29 | 0 | 33 |
| 30 | 1 | 34 |
| 31 | 1 | 35 |
| 32 | 0 | 35 |
| 33 | 1 | 36 |
| 34 | 1 | 37 |
| 35 | 0 | 37 |
| 36 | 0 | 37 |
| 37 | 1 | 38 |
| 38 | 1 | 39 |
| 39 | 2 | 41 |
| 40 | 1 | 42 |

| period | no of arrivals | cum arrivals |
|---|---|---|
| 1 | 0 | 0 |
| 2 | 0 | 0 |
| 3 | 1 | 1 |
| 4 | 0 | 1 |
| 5 | 2 | 3 |
| 6 | 0 | 3 |
| 7 | 1 | 4 |
| 8 | 1 | 5 |
| 9 | 1 | 6 |
| 10 | 1 | 7 |
| 11 | 1 | 8 |
| 12 | 0 | 8 |
| 13 | 1 | 9 |
| 14 | 2 | 11 |
| 15 | 1 | 12 |
| 16 | 1 | 13 |
| 17 | 1 | 14 |
| 18 | 3 | 17 |
| 19 | 0 | 17 |
| 20 | 2 | 19 |
| 21 | 0 | 19 |
| 22 | 3 | 22 |
| 23 | 1 | 23 |
| 24 | 0 | 23 |
| 25 | 1 | 24 |
| 26 | 0 | 24 |
| 27 | 1 | 25 |
| 28 | 1 | 26 |
| 29 | 0 | 26 |
| 30 | 2 | 28 |
| 31 | 2 | 30 |
| 32 | 0 | 30 |
| 33 | 0 | 30 |
| 34 | 4 | 34 |
| 35 | 0 | 34 |
| 36 | 0 | 34 |
| 37 | 0 | 34 |
| 38 | 0 | 34 |
| 39 | 2 | 36 |
| 40 | 3 | 39 |

| period | no of arrivals | cum arrivals |
|---|---|---|
| 1 | 3 | 3 |
| 2 | 1 | 4 |
| 3 | 3 | 7 |
| 4 | 1 | 8 |
| 5 | 2 | 10 |
| 6 | 0 | 10 |
| 7 | 1 | 11 |
| 8 | 1 | 12 |
| 9 | 1 | 13 |
| 10 | 0 | 13 |
| 11 | 1 | 14 |
| 12 | 4 | 18 |
| 13 | 1 | 19 |
| 14 | 1 | 20 |
| 15 | 0 | 20 |
| 16 | 0 | 20 |
| 17 | 3 | 23 |
| 18 | 1 | 24 |
| 19 | 1 | 25 |
| 20 | 0 | 25 |
| 21 | 0 | 25 |
| 22 | 2 | 27 |
| 23 | 0 | 27 |
| 24 | 2 | 29 |
| 25 | 1 | 30 |
| 26 | 0 | 30 |
| 27 | 1 | 31 |
| 28 | 1 | 32 |
| 29 | 1 | 33 |
| 30 | 1 | 34 |
| 31 | 1 | 35 |
| 32 | 0 | 35 |
| 33 | 2 | 37 |
| 34 | 1 | 38 |
| 35 | 0 | 38 |
| 36 | 0 | 38 |
| 37 | 0 | 38 |
| 38 | 1 | 39 |
| 39 | 1 | 40 |
| 40 | 1 | 41 |

| period | no of arrivals | cum arrivals |
|---|---|---|
| 1 | 1 | 1 |
| 2 | 2 | 3 |
| 3 | 3 | 6 |
| 4 | 0 | 6 |
| 5 | 2 | 8 |
| 6 | 3 | 11 |
| 7 | 0 | 11 |
| 8 | 0 | 11 |
| 9 | 0 | 11 |
| 10 | 1 | 12 |
| 11 | 1 | 13 |
| 12 | 0 | 13 |
| 13 | 0 | 13 |
| 14 | 1 | 14 |
| 15 | 0 | 14 |
| 16 | 0 | 14 |
| 17 | 1 | 15 |
| 18 | 1 | 16 |
| 19 | 0 | 16 |
| 20 | 0 | 16 |
| 21 | 0 | 16 |
| 22 | 0 | 16 |
| 23 | 1 | 17 |
| 24 | 0 | 17 |
| 25 | 0 | 17 |
| 26 | 0 | 17 |
| 27 | 1 | 18 |
| 28 | 1 | 19 |
| 29 | 0 | 19 |
| 30 | 0 | 19 |
| 31 | 2 | 21 |
| 32 | 0 | 21 |
| 33 | 3 | 24 |
| 34 | 0 | 24 |
| 35 | 1 | 25 |
| 36 | 5 | 30 |
| 37 | 1 | 31 |
| 38 | 0 | 31 |
| 39 | 1 | 32 |
| 40 | 0 | 32 |

| period | no of arrivals | cum arrivals |
|---|---|---|
| 1 | 0 | 0 |
| 2 | 1 | 1 |
| 3 | 2 | 3 |
| 4 | 2 | 5 |
| 5 | 1 | 6 |
| 6 | 1 | 7 |
| 7 | 3 | 10 |
| 8 | 1 | 11 |
| 9 | 1 | 12 |
| 10 | 0 | 12 |
| 11 | 3 | 15 |
| 12 | 1 | 16 |
| 13 | 1 | 17 |
| 14 | 0 | 17 |
| 15 | 0 | 17 |
| 16 | 2 | 19 |
| 17 | 2 | 21 |
| 18 | 1 | 22 |
| 19 | 0 | 22 |
| 20 | 1 | 23 |
| 21 | 3 | 26 |
| 22 | 1 | 27 |
| 23 | 0 | 27 |
| 24 | 2 | 29 |
| 25 | 1 | 30 |
| 26 | 0 | 30 |
| 27 | 1 | 31 |
| 28 | 1 | 32 |
| 29 | 0 | 32 |
| 30 | 0 | 32 |
| 31 | 1 | 33 |
| 32 | 1 | 34 |
| 33 | 3 | 37 |
| 34 | 2 | 39 |
| 35 | 0 | 39 |
| 36 | 2 | 41 |
| 37 | 1 | 42 |
| 38 | 1 | 43 |
| 39 | 1 | 44 |
| 40 | 1 | 45 |

The Lean Games and Simulations Book: Second Edition. Copyright © John Bicheno, 2014

# Poisson Distribution for 2 arrivals per period: 5 Games each of 40 periods.

**Game 1**

| period | no of arrivals | cum arrivals |
|---|---|---|
| 1 | 2 | 2 |
| 2 | 0 | 2 |
| 3 | 1 | 3 |
| 4 | 1 | 4 |
| 5 | 3 | 7 |
| 6 | 3 | 10 |
| 7 | 0 | 10 |
| 8 | 2 | 12 |
| 9 | 0 | 12 |
| 10 | 3 | 15 |
| 11 | 0 | 15 |
| 12 | 2 | 17 |
| 13 | 0 | 17 |
| 14 | 3 | 20 |
| 15 | 2 | 22 |
| 16 | 4 | 26 |
| 17 | 3 | 29 |
| 18 | 2 | 31 |
| 19 | 3 | 34 |
| 20 | 1 | 35 |
| 21 | 2 | 37 |
| 22 | 2 | 39 |
| 23 | 1 | 40 |
| 24 | 0 | 40 |
| 25 | 4 | 44 |
| 26 | 3 | 47 |
| 27 | 1 | 48 |
| 28 | 3 | 51 |
| 29 | 1 | 52 |
| 30 | 1 | 53 |
| 31 | 5 | 58 |
| 32 | 2 | 60 |
| 33 | 1 | 61 |
| 34 | 1 | 62 |
| 35 | 3 | 65 |
| 36 | 0 | 65 |
| 37 | 2 | 67 |
| 38 | 1 | 68 |
| 39 | 2 | 70 |
| 40 | 2 | 72 |

**Game 2**

| period | no of arrivals | cum arrivals |
|---|---|---|
| 1 | 2 | 2 |
| 2 | 2 | 4 |
| 3 | 1 | 5 |
| 4 | 0 | 5 |
| 5 | 2 | 7 |
| 6 | 2 | 9 |
| 7 | 1 | 10 |
| 8 | 5 | 15 |
| 9 | 1 | 16 |
| 10 | 1 | 17 |
| 11 | 1 | 18 |
| 12 | 3 | 21 |
| 13 | 2 | 23 |
| 14 | 0 | 23 |
| 15 | 5 | 28 |
| 16 | 1 | 29 |
| 17 | 0 | 29 |
| 18 | 2 | 31 |
| 19 | 5 | 36 |
| 20 | 2 | 38 |
| 21 | 2 | 40 |
| 22 | 1 | 41 |
| 23 | 1 | 42 |
| 24 | 1 | 43 |
| 25 | 1 | 44 |
| 26 | 3 | 47 |
| 27 | 2 | 49 |
| 28 | 2 | 51 |
| 29 | 3 | 54 |
| 30 | 3 | 57 |
| 31 | 3 | 60 |
| 32 | 1 | 61 |
| 33 | 5 | 66 |
| 34 | 3 | 69 |
| 35 | 4 | 73 |
| 36 | 2 | 75 |
| 37 | 2 | 77 |
| 38 | 3 | 80 |
| 39 | 4 | 84 |
| 40 | 2 | 86 |

**Game 3**

| period | no of arrivals | cum arrivals |
|---|---|---|
| 1 | 1 | 1 |
| 2 | 1 | 2 |
| 3 | 1 | 3 |
| 4 | 0 | 3 |
| 5 | 2 | 5 |
| 6 | 4 | 9 |
| 7 | 4 | 13 |
| 8 | 2 | 15 |
| 9 | 3 | 18 |
| 10 | 1 | 19 |
| 11 | 2 | 21 |
| 12 | 1 | 22 |
| 13 | 0 | 22 |
| 14 | 4 | 26 |
| 15 | 2 | 28 |
| 16 | 2 | 30 |
| 17 | 0 | 30 |
| 18 | 3 | 33 |
| 19 | 4 | 37 |
| 20 | 2 | 39 |
| 21 | 1 | 40 |
| 22 | 3 | 43 |
| 23 | 3 | 46 |
| 24 | 1 | 47 |
| 25 | 2 | 49 |
| 26 | 1 | 50 |
| 27 | 1 | 51 |
| 28 | 2 | 53 |
| 29 | 2 | 55 |
| 30 | 1 | 56 |
| 31 | 5 | 61 |
| 32 | 0 | 61 |
| 33 | 0 | 61 |
| 34 | 2 | 63 |
| 35 | 2 | 65 |
| 36 | 1 | 66 |
| 37 | 0 | 66 |
| 38 | 1 | 67 |
| 39 | 1 | 68 |
| 40 | 0 | 68 |

**Game 4**

| period | no of arrivals | cum arrivals |
|---|---|---|
| 1 | 3 | 3 |
| 2 | 1 | 4 |
| 3 | 1 | 5 |
| 4 | 2 | 7 |
| 5 | 6 | 13 |
| 6 | 0 | 13 |
| 7 | 2 | 15 |
| 8 | 3 | 18 |
| 9 | 3 | 21 |
| 10 | 3 | 24 |
| 11 | 3 | 27 |
| 12 | 2 | 29 |
| 13 | 2 | 31 |
| 14 | 3 | 34 |
| 15 | 3 | 37 |
| 16 | 4 | 41 |
| 17 | 1 | 42 |
| 18 | 0 | 42 |
| 19 | 4 | 46 |
| 20 | 3 | 49 |
| 21 | 0 | 49 |
| 22 | 2 | 51 |
| 23 | 3 | 54 |
| 24 | 1 | 55 |
| 25 | 1 | 56 |
| 26 | 2 | 58 |
| 27 | 0 | 58 |
| 28 | 1 | 59 |
| 29 | 2 | 61 |
| 30 | 5 | 66 |
| 31 | 3 | 69 |
| 32 | 0 | 69 |
| 33 | 2 | 71 |
| 34 | 2 | 73 |
| 35 | 1 | 74 |
| 36 | 1 | 75 |
| 37 | 1 | 76 |
| 38 | 2 | 78 |
| 39 | 1 | 79 |
| 40 | 2 | 81 |

**Game 5**

| period | no of arrivals | cum arrivals |
|---|---|---|
| 1 | 2 | 2 |
| 2 | 2 | 4 |
| 3 | 3 | 7 |
| 4 | 2 | 9 |
| 5 | 5 | 14 |
| 6 | 1 | 15 |
| 7 | 4 | 19 |
| 8 | 5 | 24 |
| 9 | 2 | 26 |
| 10 | 3 | 29 |
| 11 | 2 | 31 |
| 12 | 1 | 32 |
| 13 | 0 | 32 |
| 14 | 3 | 35 |
| 15 | 3 | 38 |
| 16 | 0 | 38 |
| 17 | 0 | 38 |
| 18 | 3 | 41 |
| 19 | 0 | 41 |
| 20 | 2 | 43 |
| 21 | 2 | 45 |
| 22 | 1 | 46 |
| 23 | 3 | 49 |
| 24 | 3 | 52 |
| 25 | 2 | 54 |
| 26 | 3 | 57 |
| 27 | 4 | 61 |
| 28 | 1 | 62 |
| 29 | 2 | 64 |
| 30 | 1 | 65 |
| 31 | 1 | 66 |
| 32 | 5 | 71 |
| 33 | 2 | 73 |
| 34 | 4 | 77 |
| 35 | 1 | 78 |
| 36 | 2 | 80 |
| 37 | 4 | 84 |
| 38 | 3 | 87 |
| 39 | 1 | 88 |
| 40 | 4 | 92 |

# Poisson Distribution for 3 arrivals per period: 5 Games each of 40 periods.

| period | no of arrivals | cum arrivals | no of arrivals | cum arrivals | no of arrivals | cum arrivals | no of arrivals | cum arrivals | no of arrivals | cum arrivals |
|---|---|---|---|---|---|---|---|---|---|---|
| 1 | 3 | 3 | 2 | 2 | 4 | 4 | 4 | 4 | 3 | 3 |
| 2 | 3 | 6 | 2 | 4 | 1 | 5 | 6 | 10 | 3 | 6 |
| 3 | 2 | 8 | 3 | 7 | 6 | 11 | 3 | 13 | 1 | 7 |
| 4 | 3 | 11 | 6 | 13 | 3 | 14 | 7 | 20 | 2 | 9 |
| 5 | 0 | 11 | 3 | 16 | 0 | 14 | 4 | 24 | 3 | 12 |
| 6 | 2 | 13 | 6 | 22 | 5 | 19 | 3 | 27 | 5 | 17 |
| 7 | 2 | 15 | 5 | 27 | 6 | 25 | 4 | 31 | 6 | 23 |
| 8 | 2 | 17 | 2 | 29 | 7 | 32 | 4 | 35 | 3 | 26 |
| 9 | 4 | 21 | 6 | 35 | 4 | 36 | 1 | 36 | 2 | 28 |
| 10 | 0 | 21 | 5 | 40 | 3 | 39 | 3 | 39 | 3 | 31 |
| 11 | 4 | 25 | 2 | 42 | 3 | 42 | 3 | 42 | 3 | 34 |
| 12 | 1 | 26 | 2 | 44 | 1 | 43 | 7 | 49 | 3 | 37 |
| 13 | 3 | 29 | 1 | 45 | 3 | 46 | 3 | 52 | 3 | 40 |
| 14 | 3 | 32 | 3 | 48 | 3 | 49 | 2 | 54 | 2 | 42 |
| 15 | 1 | 33 | 3 | 51 | 1 | 50 | 5 | 59 | 0 | 42 |
| 16 | 2 | 35 | 4 | 55 | 3 | 53 | 3 | 62 | 4 | 46 |
| 17 | 3 | 38 | 1 | 56 | 4 | 57 | 1 | 63 | 5 | 51 |
| 18 | 3 | 41 | 4 | 60 | 4 | 61 | 4 | 67 | 1 | 52 |
| 19 | 4 | 45 | 2 | 62 | 6 | 67 | 2 | 69 | 1 | 53 |
| 20 | 2 | 47 | 2 | 64 | 2 | 69 | 4 | 73 | 1 | 54 |
| 21 | 4 | 51 | 3 | 67 | 5 | 74 | 1 | 74 | 2 | 56 |
| 22 | 4 | 55 | 2 | 69 | 11 | 85 | 2 | 76 | 1 | 57 |
| 23 | 8 | 63 | 6 | 75 | 3 | 88 | 1 | 77 | 4 | 61 |
| 24 | 1 | 64 | 2 | 77 | 3 | 91 | 3 | 80 | 2 | 63 |
| 25 | 2 | 66 | 5 | 82 | 2 | 93 | 8 | 88 | 8 | 71 |
| 26 | 1 | 67 | 1 | 83 | 3 | 96 | 2 | 90 | 3 | 74 |
| 27 | 3 | 70 | 1 | 84 | 2 | 98 | 3 | 93 | 4 | 78 |
| 28 | 1 | 71 | 4 | 88 | 3 | 101 | 5 | 98 | 1 | 79 |
| 29 | 5 | 76 | 2 | 90 | 2 | 103 | 3 | 101 | 2 | 81 |
| 30 | 1 | 77 | 2 | 92 | 3 | 106 | 1 | 102 | 2 | 83 |
| 31 | 5 | 82 | 2 | 94 | 1 | 107 | 2 | 104 | 6 | 89 |
| 32 | 2 | 84 | 6 | 100 | 3 | 110 | 3 | 107 | 2 | 91 |
| 33 | 1 | 85 | 2 | 102 | 2 | 112 | 2 | 109 | 9 | 100 |
| 34 | 4 | 89 | 1 | 103 | 3 | 115 | 4 | 113 | 5 | 105 |
| 35 | 3 | 92 | 2 | 105 | 3 | 118 | 1 | 114 | 0 | 105 |
| 36 | 2 | 94 | 3 | 108 | 6 | 124 | 1 | 115 | 3 | 108 |
| 37 | 5 | 99 | 2 | 110 | 3 | 127 | 1 | 116 | 3 | 111 |
| 38 | 6 | 105 | 0 | 110 | 4 | 131 | 3 | 119 | 4 | 115 |
| 39 | 4 | 109 | 6 | 116 | 3 | 134 | 3 | 122 | 4 | 119 |
| 40 | 5 | 114 | 3 | 119 | 2 | 136 | 0 | 122 | 5 | 124 |

The Lean Games and Simulations Book: Second Edition. Copyright © John Bicheno, 2014

# Rolls to obtain a 6
## ( Two Typical Results)

**Chart 1** — Here, average no of rolls to get a 6 is 4.9

| | 1 | 2 | 3 | 4 | 5 | 6 | 7 | 8 | 9 | 10 | 11 | 12 |
|---|---|---|---|---|---|---|---|---|---|---|---|---|
| | | | | | X | | | | | | | |
| | | | | | X | | | | | | | |
| | | X | | | X | | | | | | | |
| | X | X | X | | X | X | X | | | | | |
| | X | X | X | | X | X | X | X | | X | X | |
| No | 1 | 2 | 3 | 4 | 5 | 6 | 7 | 8 | 9 | 10 | 11 | 12 |

**Chart 2** — Here, average no of rolls to get a 6 is 5.5

| | 1 | 2 | 3 | 4 | 5 | 6 | 7 | 8 | 9 | 10 | 11 | 12 | 13 | 14 |
|---|---|---|---|---|---|---|---|---|---|---|---|---|---|---|
| | | | X | | | | | | | | | | | |
| | | | X | | | | X | | | | | | | |
| | X | X | X | X | X | X | X | X | | X | | | X | |
| | X | X | X | X | X | X | X | X | | X | | X | X | X |
| No | 1 | 2 | 3 | 4 | 5 | 6 | 7 | 8 | 9 | 10 | 11 | 12 | 13 | 14 |

The Lean Games and Simulations Book: Second Edition. Copyright © John Bicheno, 2014

# Service Call Centre Game

## Typical Results for 1 arrival per Period (Poisson)

(Note: In this example 6 is taken as the average number of rolls to get a 6; with 2 rolls per period it takes an average of 3 periods to service a customer. Hence 3 operators are utilized 100% on average.)

| Operators | Max Queue | Ave Queue | Utiliz | Abandoned* | Server Cost | Wait Cost | Total Cost | |
|---|---|---|---|---|---|---|---|---|
| 3 | 11 | 2.9 | 100% | 7 | £300 | £0 | £300 | Waiting is assumed to be free (!) |
| | | | | | £300 | £290 | £590 | Waiting at £2.5 per period |
| | | | | | £300 | £1160 | £1460 | Waiting at £10 per period |
| 4 | 3 | 0.8 | 75% | 8 | £400 | £0 | £400 | Free |
| | | | | | £400 | £80 | £480 | Waiting at £2.5 per period |
| | | | | | £400 | £320 | £720 | Waiting at £10 per period |
| 5 | 3 | 0.2 | 60% | 6 | £500 | £0 | £500 | Free |
| | | | | | £500 | £20 | £520 | Waiting at £2.5 per period |
| | | | | | £500 | £80 | £580 | Waiting at £10 per period |

* Calls abandoned when target is set at 8 rolls

# Call Centre Game: Operator Instructions
(one of these sheets for each operator)

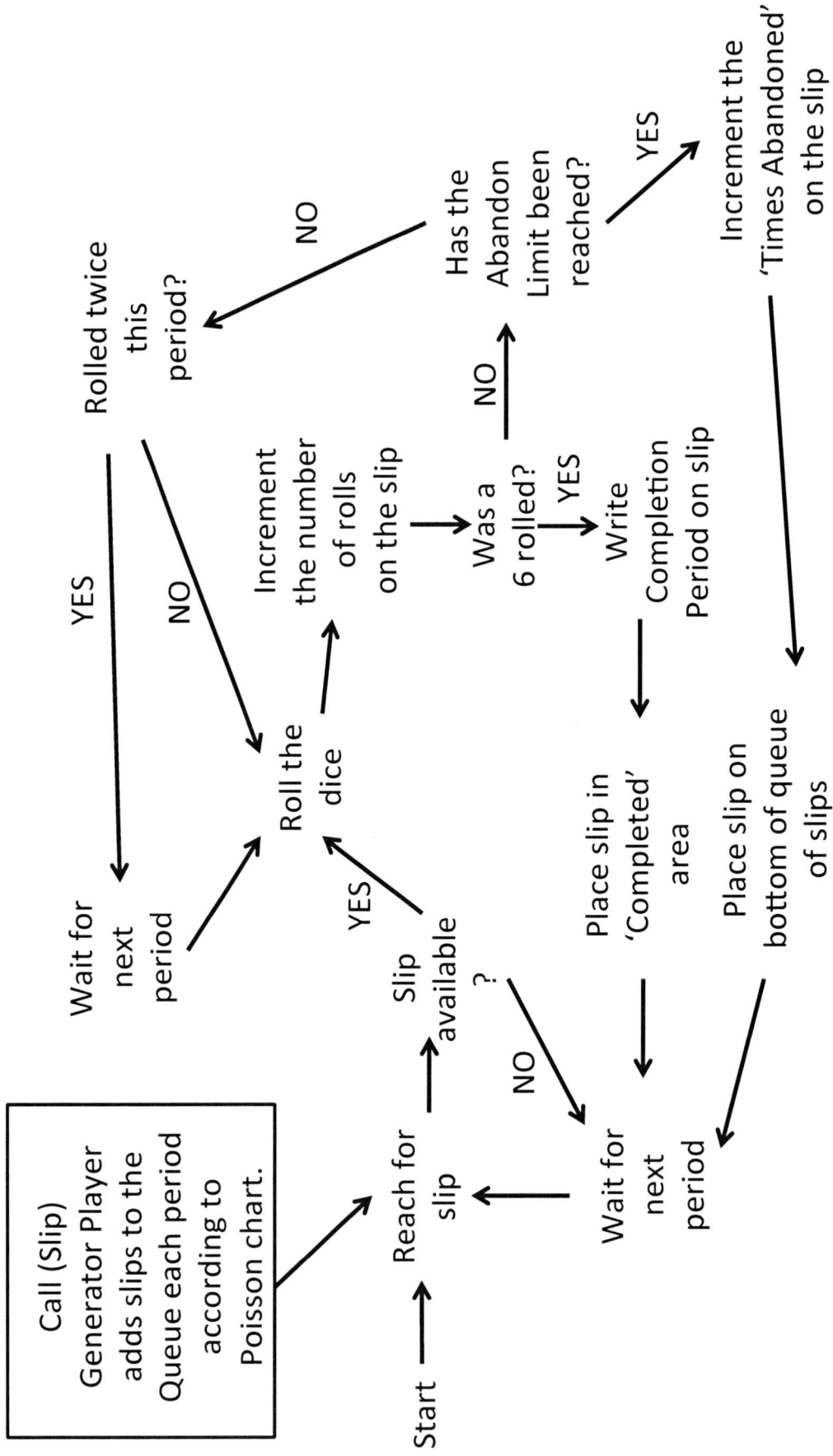

```
Call (Slip)
Generator Player
adds slips to the
Queue each period
according to
Poisson chart.
```

Start → Reach for slip

Reach for slip → Slip available?

Slip available? — YES → Roll the dice

Slip available? — NO → Wait for next period

Wait for next period → Reach for slip

Roll the dice → Increment the number of rolls on the slip

Increment the number of rolls on the slip → Was a 6 rolled?

Was a 6 rolled? — YES → Write Completion Period on slip

Was a 6 rolled? — NO → Has the Abandon Limit been reached?

Write Completion Period on slip → Place slip in 'Completed' area

Has the Abandon Limit been reached? — YES → Increment the 'Times Abandoned' on the slip

Has the Abandon Limit been reached? — NO → Rolled twice this period?

Rolled twice this period? — YES → Wait for next period

Rolled twice this period? — NO → Roll the dice

Increment the 'Times Abandoned' on the slip → Place slip on bottom of queue of slips

Place slip in 'Completed' area → Wait for next period

Place slip on bottom of queue of slips → Wait for next period

# Call and Service Centre Game

## Customer Call Slips: Cut out 80 per game

| | |
|---|---|
| Customer No:<br>Abandonment limit:<br>Arrival period:<br>Completion period:<br>Times abandoned:<br>No of rolls: | Customer No:<br>Abandonment limit:<br>Arrival period:<br>Completion period:<br>Times abandoned:<br>No of rolls: |
| Customer No:<br>Abandonment limit:<br>Arrival period:<br>Completion period:<br>Times abandoned:<br>No of rolls: | Customer No:<br>Abandonment limit:<br>Arrival period:<br>Completion period:<br>Times abandoned:<br>No of rolls: |
| Customer No:<br>Abandonment limit:<br>Arrival period:<br>Completion period:<br>Times abandoned:<br>No of rolls: | Customer No:<br>Abandonment limit:<br>Arrival period:<br>Completion period:<br>Times abandoned:<br>No of rolls: |
| Customer No:<br>Abandonment limit:<br>Arrival period:<br>Completion period:<br>Times abandoned:<br>No of rolls: | Customer No:<br>Abandonment limit:<br>Arrival period:<br>Completion period:<br>Times abandoned:<br>No of rolls: |

# Spot the Rot

(Acknowledgement to Peter Willmott of Willmott Solutions)

- Go to a designated machine or area and list 25 things that are not absolutely perfect

- Don't stop the machine

- Observe carefully: top down, bottom up, in front, behind, underneath, obscured

- This could include the supporting area, floor, tools, spares, lighting, connections, signage

- Also how clear is the schedule, defects, stoppage, warnings, standards / SOPS

# TV News Game

# Cross Functional Team Exercise: The Daily News

- You are a TV news production company.
- You have 1 hour to make a 5 minute newscast.
- The newscast must include news items and a short documentary
- There are 5 departments
  - Editorial
  - Research
  - Writing
  - Production
  - News Readers
- No internet, printers, copiers.
- Laptop computers or iPads are allowed but no internet.

The Original game was developed by Kate Mackie of Thinkflow

# The News Game (2)

- The stories must be
  - Real
  - Topical
  - Balance not biased
  - Factual
  - Contain at least 3 quantitative facts
  - Interesting
- The documentary must include visuals (e.g. diagrams, drawings, graphs) and must be approved for quality.
- There are two news readers

# The News Game (3) Editorial

- You choose the news stories and the documentary
- You decide the time split between news and documentary
- You tell Research what to research – in writing.
- You do not have any input with any other department
- Once you have decided on the stories you will then and only then communicate with the Research team – once only!
- You must approve the final documentary before broadcast.
- Editors may discuss Production requirements with the Production section.
- Editors may monitor progress but may not discuss progress with individual sections or staff.

# The News Game (6) Production

- Your task is to set up the room so that that everyone can watch the news reports being presented.

- You must explain to the Reading team how and when and where the stories are to be presented.

- You receive the stories from the Writing Team and pass them on to the Reading Team.

- Some text editing is allowed.

- You only communicate with each of these teams once

- You do not communicate with any other team.

- The newscast will begin in EXACTLY one hour.

# The News Game (7) News Reading Team

- You must read out the news stories you receive from the Production Team

- You only communicate with the Production Team once

- You do not communicate with any other team.

- When you are ready to present you must sit at the news desks and announce that you are ready

- You will then present your news reports to the group.

- Your newscast will take exactly 5 minutes.

# The News Game (8)

- Was this Lean?
- What about Cost, Quality, Delivery, Time?
- What about value adding?
- What about communication?
- Who was the customer
- Who was the final customer
- What about specification?

# The News Game (9): Second Round

- The Team (Crew) will now have 15 minutes to decide on how to reorganise for a second 5 minute newscast.

- The Editors and News readers remain. Other functions remain, but may be reorganised.

- The Team will then be given 20 minutes to prepare for a second 5-minute news cast.

- New stories and documentary must be chosen.

- No communication restrictions.

- Once again, no internet, printers, copiers. Laptop computers are allowed.

# Features of Toyota Design Practice

- Role of Chief Engineer
- Authority of Functional Managers
- Check Sheets
- Leveling the work
- 'Cadence' (the drum beat of moving forward)
- Bookshelving
- 'Set Based' Design
- The 'Obeya' (Big) Room

# Other Considerations: Mascitelli's Rules

1. No more than two major projects per developer
2. Among the projects, priorities must be made clear
3. Capacity – avoid overload
4. Allocate time in 'time slices'
5. Time slices should be a minimum duration (2 hours in a real project)
6. Critical activities should enjoy priority
7. All 'walk in' demands must pass through a priority filter
8. Workflow should be managed visually

Adapted from Ronal Mascitelli, *Mastering Lean Product Development*, Technology Perspectives, 2011

# Features of Goldratt's Critical Chain Project Management

- The Critical Chain rather than the Critical Path

- Time slack at the end, for the whole project, but not for individual activities

- Time buffers at convergent paths, to prevent non-critical holding up critical activities

# Features of 'Lean Construction' Project Management

- Develop and grow check lists covering every requirement for every activity. Check these off in advance.

- Not just monitor the critical path, but monitor the consumption of the total time buffer.

- 'Last Planner' segregates work into weekly work plans that have three characteristics:

  1. Work is selected in the right order in line with project deliverables.

  2. The amount of work selected is achievable in the timescale

  3. The work selected has no constraints preventing it from being completed

# Multitasking

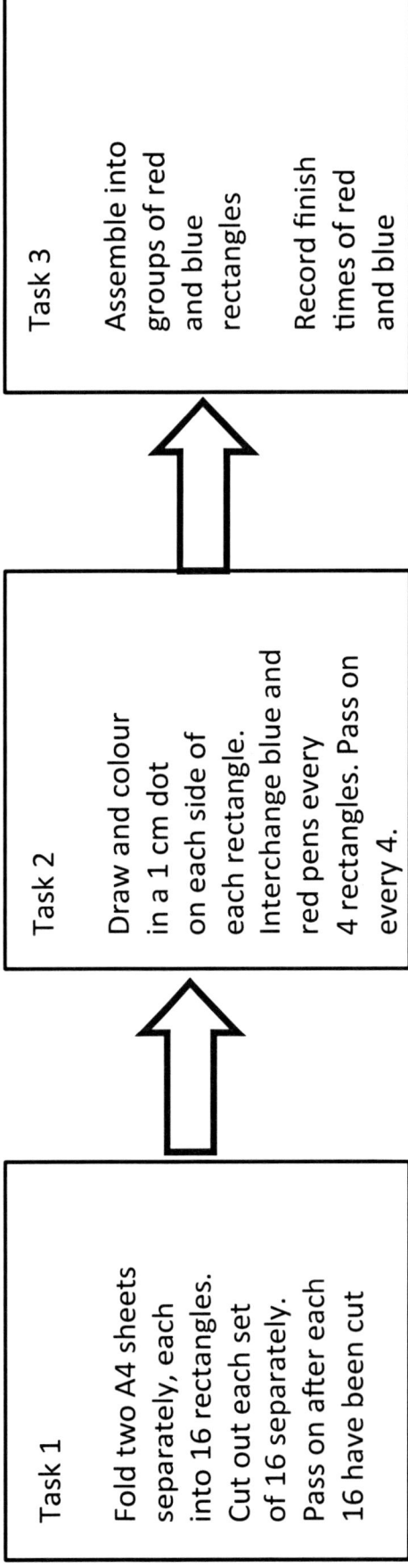

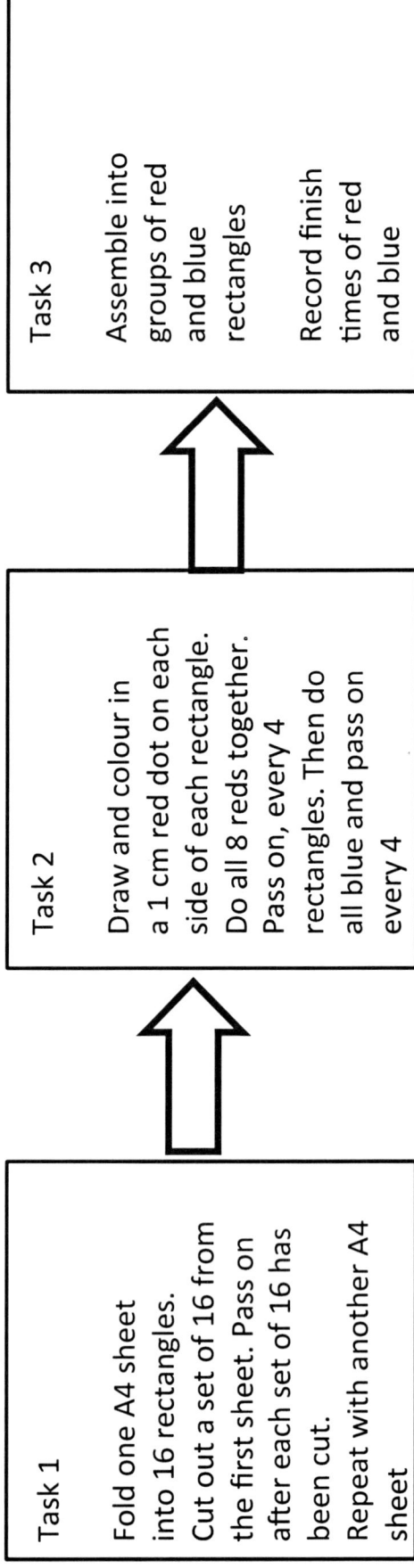

# Multitasking Game Sheet

## Round 1

**Task 1**

Fold two A4 sheets separately, each into 16 rectangles. Cut out each set of 16 separately. Pass on after each 16 have been cut

**Task 2**

Draw and colour in a 1 cm dot on each side of each rectangle. Interchange blue and red pens every 4 rectangles. Pass on every 4.

**Task 3**

Assemble into groups of red and blue rectangles

Record finish times of red and blue

## Round 2

**Task 1**

Fold one A4 sheet into 16 rectangles. Cut out a set of 16 from the first sheet. Pass on after each set of 16 has been cut. Repeat with another A4 sheet

**Task 2**

Draw and colour in a 1 cm red dot on each side of each rectangle. Do all 8 reds together. Pass on, every 4 rectangles. Then do all blue and pass on every 4

**Task 3**

Assemble into groups of red and blue rectangles

Record finish times of red and blue

The Lean Games and Simulations Book: Second Edition. Copyright © John Bicheno, 2014

# Multitasking Instructions

- Two rounds

- Several parallel games

- In Round 1 each game should have 5 players: Three people doing the tasks and two 'managers'

- The managers are interested in their separate red and blue projects. They only allow 4 of the other colour to be made before changing to their colour. They shout encouragement

- In round 2, no managers are needed. The players complete all red rectangles before starting on blue

- Compare the results

This game has been adapted from a game in Gonzalez-Rivas and Larsson 'Far From The Factory', CRC Press, 2011. See the excellent notes in Chapter 4

# Simultaneous Project Working

- In many design and project offices it is usual for team members to work on several projects simultaneously.

- This is because of specialist work combined with interdependency.

- You or your group may have to wait for completion or approval of a previous stage.

- If there is one project at a time, groups may be idle waiting for the previous stage.

- But how many simultaneous projects should be taken on at the same time?

- This game or exercise explores this.

# The Projects

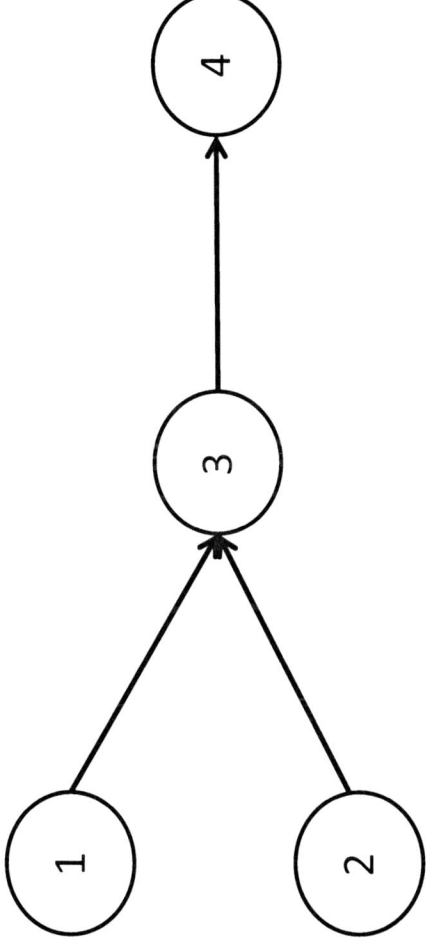

Notes on the projects

- Each project has only 4 activities, with 4 players, one player per activity.
- Each period, each player rolls a dice.
- The activity is completed when a 6 is rolled.
- This is a good simulation of a Beta Distribution very common in projects
- Each project has a distinct colour : Red, Blue, Green, Yellow
- Coloured discs are moved between activities
- Activities 1 and 2 each start with similar colour discs.
- Activity 3 can only begin when there are two discs (from activities 1 and 2).
- Activity 4 can only begin when activity 3 is complete

# Simultaneous Projects Game Sheets

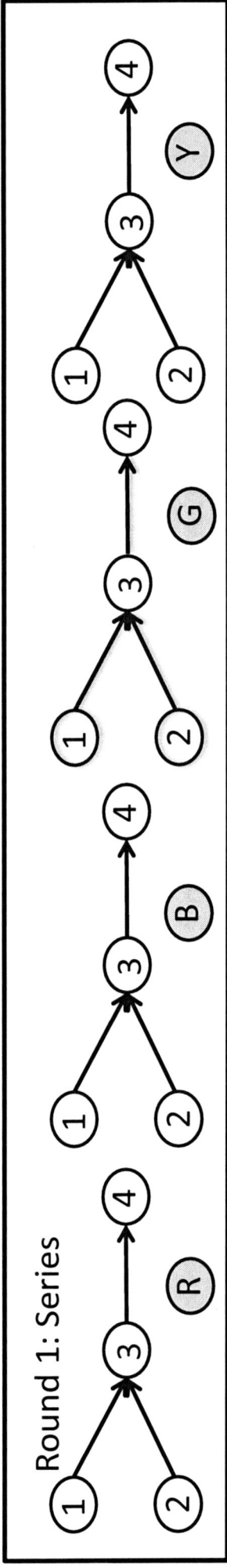

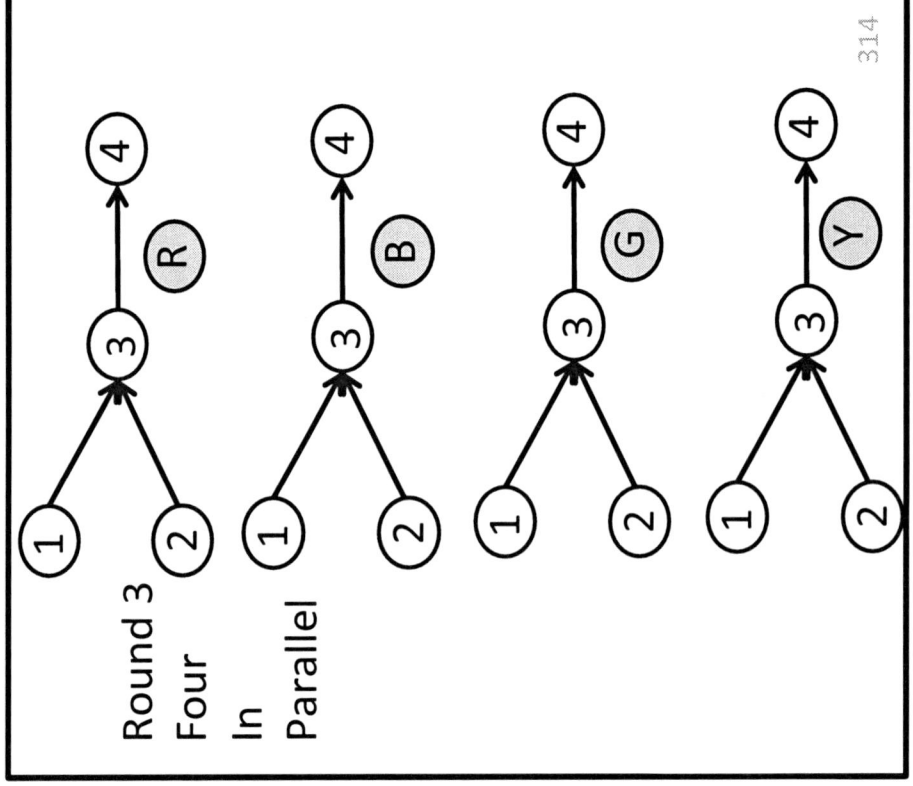

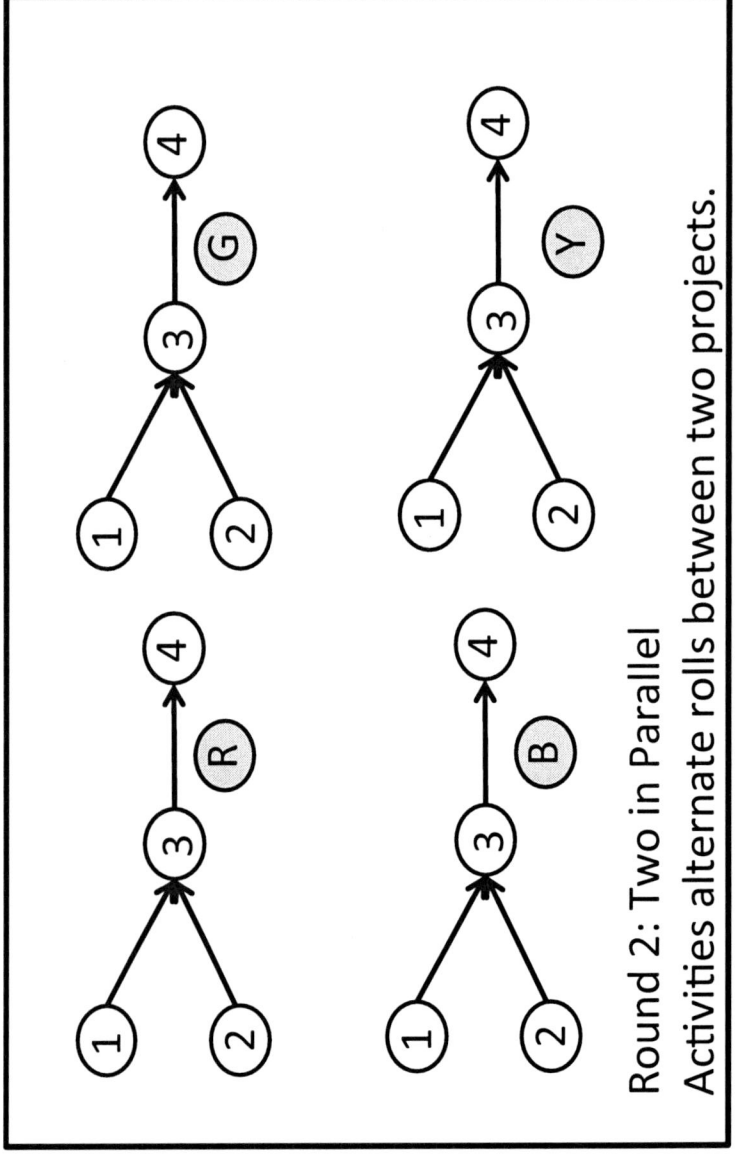

Round 1: Series

Round 2: Two in Parallel
Activities alternate rolls between two projects.

Round 3
Four
In
Parallel

# Simultaneous Projects
## Player 1 and 2 Instructions

- There are 4 projects per round (Red, Blue, Green, Yellow) and 3 rounds per game
- There will probably be several parallel games and you play one of these.
- The timing of each round will be controlled by a display showing the current period. Rolla dice once per period only.
- A project has 4 activities. You are concerned with Activity 1 (Player 1), and Activity 2 (Player 2). Since there are 4 projects you have 4 activities to complete
- You begin each round with
  - a Red disc at Activity 1 (Player 1) and at Activity 2 (Player2) for the Red project
  - A Blue disc at Activity 1 (Player 1) and at Activity 2 (Player2) for the Blue project
  - A Green disc at Activity 1 (Player 1) and at Activity 2 (Player2) for the Green project
  - A Yellow disc at Activity 1 (Player 1) and at Activity 2 (Player2) for the Yellow project
- Each period roll a dice. If you roll a 6 the current activity is complete. If you roll 1 to 5 the activity is not complete.
- The sequence of rolls is as follows:
  - Round 1: keep rolling on red until you get a 6, then begin rolling on blue until you get a 6, then begin rolling on green until you get a 6, then begin rolling on yellow until you get a 6.
  - Round 2: Alternate rolls between red and blue. When a 6 is rolled, stop rolling for that colour and begin rolling for the next colour. Continue alternating for the new pair until a 6 is rolled. Then begin the third pair until a 6 is rolled. When there is only one project remaining continue rolling each period until a 6 is rolled.
  - Round 3: Alternate between red, blue, green, and yellow. When a 6 is rolled alternate between the remaining three colours until a 6, then alternate between the remaining 2. When there is only one project remaining continue rolling each period until a 6 is rolled.
- When a 6 is rolled move the appropriately coloured disc to Activity 3

The Lean Games and Simulations Book: Second Edition. Copyright © John Bicheno, 2014

# Simultaneous Projects
# Player 3 Instructions

- There are 4 projects per round (Red, Blue, Green, Yellow) and 3 rounds per game

- There will probably be several parallel games

- The timing of each round will be controlled by a display showing the current period. Roll once per period.

- A project has 4 activities. You are concerned with Activity 3. Since there are 4 projects you have 4 activities to complete.

- You may only start rolling a dice when you have two discs of the same colour at Activity 3.

- Each period roll a dice. If you roll a 6 the current activity is complete. If you roll a 1 to 5 the activity is not complete.

- The sequence of rolls is as follows:

  - Round 1: Wait until you have two red discs. Then keep rolling on red until you get a 6, then if there are two blue discs at Activity 3 begin rolling on blue until you get a 6. Similarly for Green and then Yellow.

  - Round 2: Wait until you either have two red discs or two blue discs. Then start rolling, one roll per period. Alternate rolls between a maximum of two projects at a time. A 6 must be rolled before you can begin a third project – but you may have to wait until two discs are available. Continue alternating until a 6 is rolled. Similarly for the fourth project. When there is only one project remaining continue rolling each period until a 6 is rolled.

  - Round 3: Wait until you have two red or two blue discs or two green or two yellow discs. Then start rolling, one roll per period. Alternate rolls between all available projects. Continue alternating until a 6 is rolled, then stop rolling for that project but keep alternating between all available projects. When there is only one project remaining continue rolling each period until a 6 is rolled.

- When a 6 is rolled move the pair of appropriately coloured discs of the same colour to Activity 4

# Simultaneous Projects
# Player 4 Instructions

- There are 4 projects per round (Red, Blue, Green, Yellow) and 3 rounds per game
- There will probably be several parallel games
- The timing of each round will be controlled by a display showing the current period. Roll once per period.
- A project has 4 activities. You are concerned with Activity 4. Since there are 4 projects you have 4 activities to complete.
- You may only start rolling a dice when you have two discs of the same colour at Activity 4.
- Each period roll a dice. If you roll a 6 the current activity is complete. If you roll 1 to 5 the activity is not complete.
- The sequence of rolls is as follows:

  - Round 1: Wait until you have two red discs. Then keep rolling on red until a 6, then if there are two blue discs at Activity 4 begin rolling on blue until a 6. Similarly for Green and then Yellow. Projects must be completed in the sequence R, B, G, Y.

  - Round 2: Wait until you either have two red discs or two blue discs. Start rolling. Alternate rolls between a maximum of two projects at a time. A 6 must be rolled before you can begin a third project – but you may have to wait until two discs are available. Continue alternating until a 6 is rolled. Similarly for the fourth project. Project R must be completed before project G can begin. Project B must be completed before Project Y can begin.

  - Round 3: Wait until you have two red or two blue discs or two green or two yellow discs.  Start rolling.  Alternate rolls between all available projects. Continue alternating until a 6 is rolled, then stop rolling for that project but keep alternating between all available projects. When there is only one project remaining continue rolling each period until a 6 is rolled.

- When a 6 is rolled move the pair of appropriately coloured discs of the same colour to Activity 4

# Simultaneous Projects
# Points and Typical Results

## Points

- Sequential projects loose lots of time waiting for prior activities
- Simultaneous projects reduce waits at the cost of extended times for parallel activities.
- Too many parallel projects lead to extended project times.
- There is an optimal balance
- Maybe two or three simultaneous projects is best
- This is a simple case; real project co-ordination increases complexity exponentially!

## Typical Results: Completion Periods

- Round 1
  1. 23
  2. 42
  3. 49
  4. 52

- Round 2
  1. 11
  2. 12
  3. 15
  4. 27

- Round 3
  1. 19
  2. 21
  3. 30
  4. 34

# Simultaneous Projects Discs
## (Laminate and Cut Out is Best)

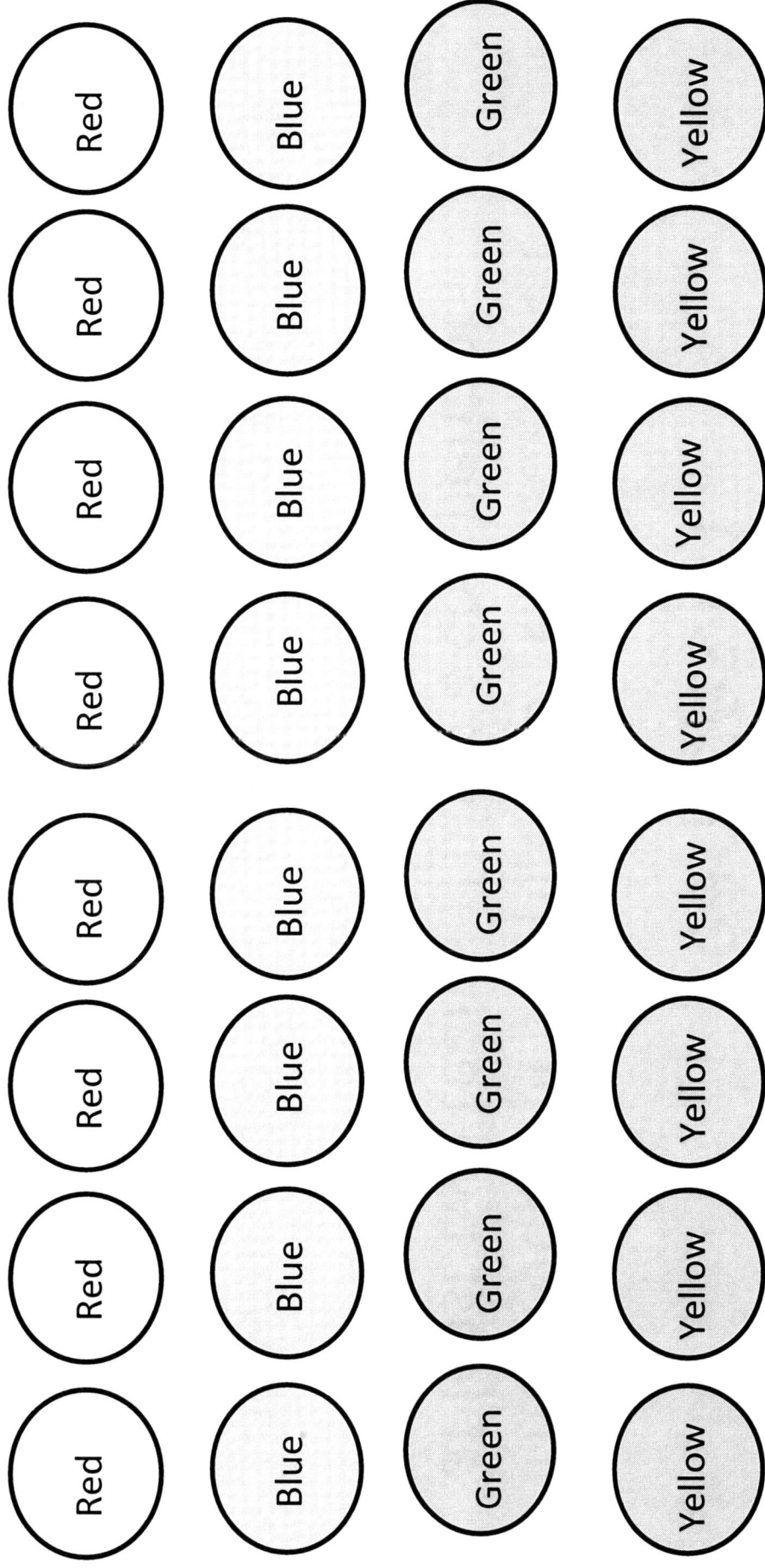

| Red | Red | Red | Red | Red | Red | Red |
| Blue | Blue | Blue | Blue | Blue | Blue | Blue |
| Green | Green | Green | Green | Green | Green | Green |
| Yellow | Yellow | Yellow | Yellow | Yellow | Yellow | Yellow |

# Lean Design

- The group will be split into Teams. There must be at least two teams. This is a competition.

- There will be two projects

- The winners of each project will be the team the makes the highest profit (or makes at the lowest cost)

- The full game takes approximately 3 hours

# Lean Design: Project 1

- Your team is to build a tower in the next 45 minutes, using the materials available.

- Materials are to be collected from the supplier and cannot be returned. Materials consist of food cans, packaging tape, A4 size cardboard, string, blue tac, scissors, rulers. (Rulers are not to be used as part of the structure).

- The final tower must be capable of withstanding a blow (puff) from the mouth of the instructor, to take place at two thirds (from base) height.

- Value of the tower is £100 per cm of height

- There is a maximum bonus of £500 for elegant design.

The Lean Games and Simulations Book: Second Edition. Copyright © John Bicheno, 2014

# Lean Design: Project 2

- Your team is to build a bridge spanning two tables 30 cm apart, during the next 60 minutes.

- Materials are to be collected from the supplier and cannot be returned.

- The bridge must be capable of supporting a small food can at midspan

- Height of lowest point above tables gains £100 per cm. Minus £100 for every cm of lowest point below table height.

- Bonus for good aesthetic design, craftsmanship, emotional appeal, elegance and sophistication, is £1000 total.

- There is a design review by the customer after 40 minutes. During this review the clock will stop. The instructor (customer) will walk around to each team and hear a 2 minute (strict) presentation.

- The design and build period may be extended at a cost of £100 per minute, up to 10 minutes

The Lean Games and Simulations Book: Second Edition. Copyright © John Bicheno, 2014

# Materials

- Food cans: £100 each (maximum 9)
- 5 cm wide packaging tape : £20 per cm (to be measured after completion. Discarded tape must be kept for measurement)
- Scissors : £100 each
- String: £5 per cm.
- Rulers: £50 each
- Cardboard: £50 per A4 size sheet
- BlueTac: £30 per slab

# Between Round 1 and Round 2
## (30 minutes)

- Your team should discuss
  - better ways of managing a design and build project (10 minutes all, 10 minutes sub-group)
  - (optional) use of Planning Board (5 minutes)
  - Alternatives for spanning a gap (20 minutes, sub-group)
- You have 30 minutes available
- Materials are free during this period
  - Limited quantities are available
  - All materials must be returned before the next round

# Questions after Round 1

- Did you ask the customer? When? How often?
- Were the criteria clear?
- Were the 'must be', 'performance factors', and 'delighters' clear?
- Did you test prototypes?
- Did you have a project plan?
- Was there a project manager?
- Do you know how time was spent?
- Did you plan the phases?
- Did you have target times for the phases?
- Were there concurrent activities? Could there have been?
- Did you use functional managers / experts?
- Set based design?
- How did you do the cost tradeoffs?
- Did you use Pugh analysis?

# Design Game: Double Diamond

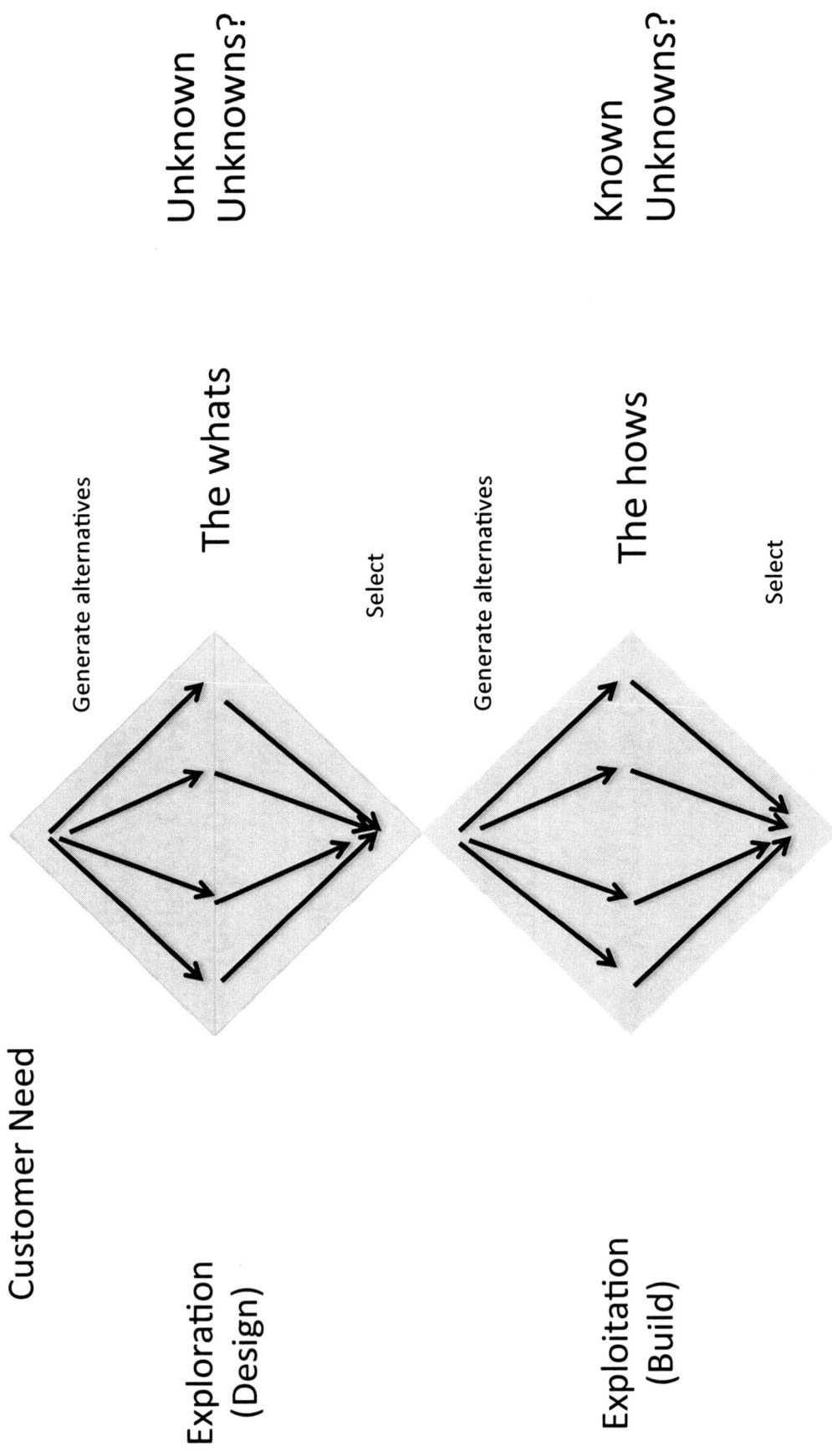

Customer Need

Generate alternatives

The whats

Select

Unknown Unknowns?

Exploration (Design)

Generate alternatives

The hows

Select

Known Unknowns?

Exploitation (Build)

The Lean Games and Simulations Book: Second Edition. Copyright © John Bicheno, 2014

# Design Planning

## (in 'Obeya' Room?)

| Activities | Time | | | | | |
|---|---|---|---|---|---|---|
| | 10 | 11 | 12 | 13 | 14 | 15 |
| Interview Customer | Kim | | | | | |
| Feasibility Study | | John B and Simon | | | | |
| Cost Studies | | | John D | | | |
| Concept Develop | | | | John B Chris | | |
| Stage Gate | | | | | All | |
| Etc | | | | | | |
| Etc | | | | | | |

The Lean Games and Simulations Book: Second Edition. Copyright © John Bicheno, 2014

# Gap Spanning Alternatives

Truss

Suspension

I Beam Plate Girder X Section

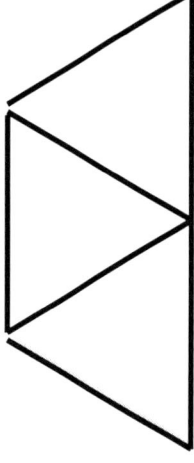

Box Beam X Section

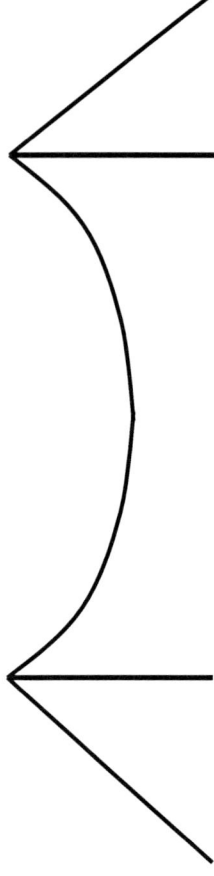

# Design Game Evaluation

## (Not to be shown before end of Round One)

| | Teams | | | | |
|---|---|---|---|---|---|
| | 1 | 2 | 3 | 4 | 5 |
| Performance and Cost | | | | | |
| Craftsmanship | | | | | |
| Emotional Appeal | | | | | |
| Elegance and Sophistication | | | | | |
| 'Green' and Environmental Aspects | | | | | |

This table has been adapted from James Adams, *Good Products, Bad Products*, McGraw Hill, 2012. Adams also uses 'Human Fit' (i.e. ergonomics), and 'Symbolism and Cultural Values' (e.g. German cars, Wine or beer culture, etc.)

# The Happy Pig

# Happy Pig

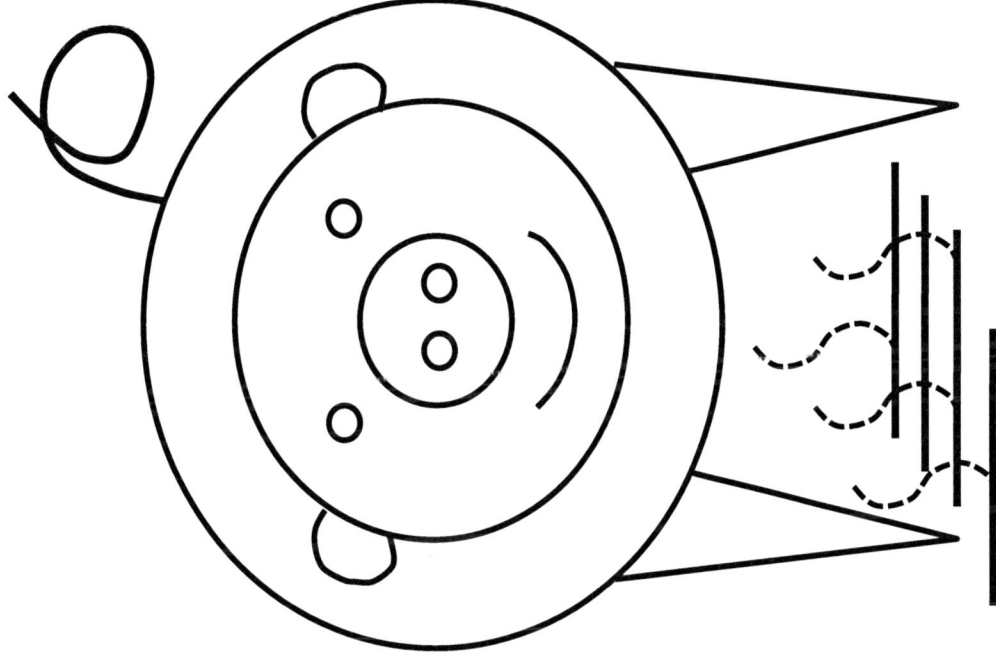

# Standards

"To standardise a method is to choose out of many methods the best one, and use it. What is the best way to do a thing? It is the sum of all the good ways we have discovered up to the present. It, therefore becomes the standard.

Today's standardisation is the necessary foundation on which tomorrow's improvement will be based. If you think of 'standardisation as the best we know today, but which is to be improved tomorrow - you get somewhere. But if you think of standards as confining, then progress stops."

Henry Ford, *Today and Tomorrow*, 1926

# Better Happy Pig?

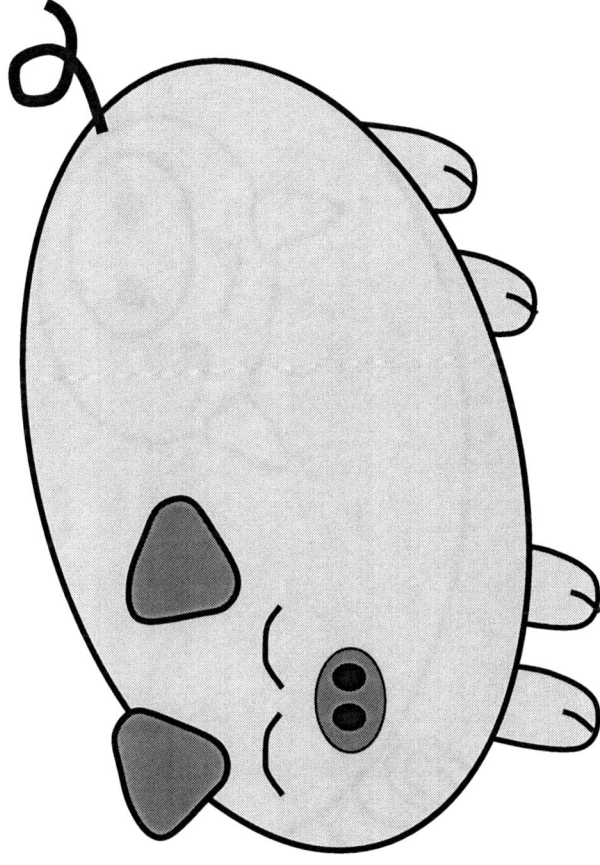

# Even Better Happy Pig?

The Lean Games and Simulations Book: Second Edition. Copyright © John Bicheno, 2014

# Drawing A Better Happy Pig?

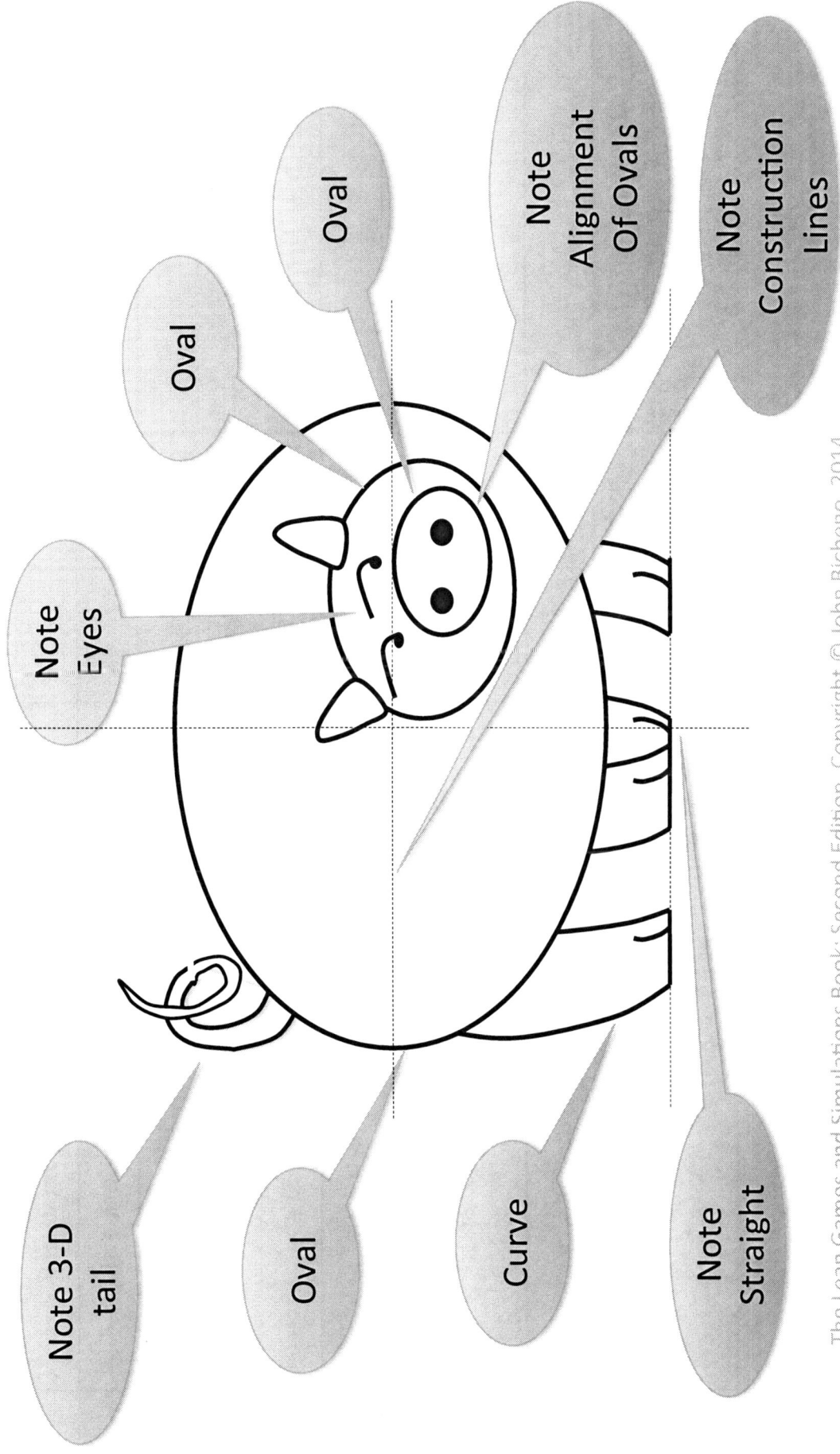

# The Standards Debate

| Job | Indicate the applicability of standard work | | | | Your comments |
|---|---|---|---|---|---|
| | Low | | High | | |
| Repetitive work; no decisions | | | | | |
| Repetitive work; several decisions | | | | | |
| New Product Design | | | | | |
| Emergency Health Care | | | | | |
| Surgeon | | | | | |
| Pilot | | | | | |
| Call Centre operator | | | | | |
| Office Work: varied | | | | | |
| Kaizen facilitator | | | | | |
| Production Manager | | | | | |

The Lean Games and Simulations Book: Second Edition. Copyright © John Bicheno, 2014

# Go See, Genchi Genbutsu

## Bicycle Exercise

- Sketch a bicycle on a sheet of paper

- Include as much detail as possible

- Rate your knowledge of bicycles on a 1 to 10 scale
  (1 = no knowledge; 10 detailed knowledge)

# Bicycle Checklist

| | Yes/No | | Yes/ No |
|---|---|---|---|
| Does it have a chain? | | Any damper on front head tube? | |
| Chain connecting crank to back wheel? | | Collars to allow steering movement? | |
| Pedals connected to crank hub by tube? | | Seat ? (Shape?) | |
| Chain connecting large diameter crank hub with smaller diameter rear hub? | | Curve on front head tube just before front wheel hub? | |
| Curved handlebar? | | Seat on extended seat tube? | |
| Double triangle frame? | | Real wheel gear sprockets? | |
| Brake grip on handlebar? | | Derailleur on back wheel? | |
| Brake assembly on front wheel? | | Turnable front head tube? (How?) | |
| Brake assembly on rear wheel? | | Rear mudguard? | |
| Frame tubes on each side of wheels? | | Reflector at rear ? (Back of seat?) | |
| Rotating pedals? | | Wheel spokes that cross? | |
| Sprockets on crank hub? | | Brake cables from handlebar to wheels? | |
| Pads on handlebars? | | Brand name on a tube? | |

Now rate your Knowledge of a Bicycle again:

| Before |
|---|
| Now |

The Lean Games and Simulations Book: Second Edition. Copyright © John Bicheno, 2014

# Targets and Measures Game

- This is a simple variant on the classic Deming Red Bead Game.

- There is no doubt that the original Red Bead Game is a superior game but this offers an easy way to appreciate some common issues with targets, measures and motivation using only dice.

- Players

  - 8 volunteers (6 workers, one inspector, one scribe) or 6 volunteers (4 workers, one inspector, one scribe)

    - pair of dice is required

- Play 4 rounds with each player rolling a pair of dice and adding to get a total for the round.

- The total is verified by the inspector and written up by the scribe.

- The manager has set a target for the sum of the two rolls at 9, for each worker. This would mean a Grand Total target of 144 for 4 workers or 216 for 6 workers.

# Targets and Measures Game: Control Limits

## Notes

- The Total is the sum of all pairs of rolls for the full game (16 pairs for 4 workers; 24 pairs for 6 workers)
- The X bar column is the total divided by 16 or 24
- The Target is an average of 9 or more.
- The UCL is the Upper Control Limit for a control chart having a particular Total score. Similarly for the Lower Control Limit
- If any individual player average exceeds this UCL, the system is 'out of control'. This would represent exceptionally good performance
- If any individual player average is between the LCL and UCL the system is 'in control' and no special action should be taken.
- If any individual player average is less than the LCL this would represent exceptionally poor performance and action should be taken
- This game illustrates 'Regression to the Mean'

### For 4 workers

| Total | X Bar (ave) | LCL | UCL |
|---|---|---|---|
| 40 | 2.50 | 0.00 | 6.72 |
| 45 | 2.81 | 0.00 | 7.21 |
| 50 | 3.13 | 0.00 | 7.69 |
| 55 | 3.44 | 0.00 | 8.14 |
| 60 | 3.75 | 0.00 | 8.57 |
| 65 | 4.06 | 0.00 | 8.98 |
| 70 | 4.38 | 0.00 | 9.38 |
| 75 | 4.69 | 0.00 | 9.76 |
| 80 | 5.00 | 0.00 | 10.12 |
| 85 | 5.31 | 0.15 | 10.47 |
| 90 | 5.63 | 0.44 | 10.81 |
| 95 | 5.94 | 0.74 | 11.13 |
| 100 | 6.25 | 1.06 | 11.44 |
| 105 | 6.56 | 1.39 | 11.74 |
| 110 | 6.88 | 1.73 | 12.02 |
| 115 | 7.19 | 2.09 | 12.28 |
| 120 | 7.50 | 2.47 | 12.53 |
| 125 | 7.81 | 2.86 | 12.77 |
| 130 | 8.13 | 3.27 | 12.98 |
| 135 | 8.44 | 3.69 | 13.19 |
| 140 | 8.75 | 4.13 | 13.37 |
| 145 | 9.06 | 4.59 | 13.53 |
| 150 | 9.38 | 5.08 | 13.67 |
| 155 | 9.69 | 5.59 | 13.79 |
| 160 | 10.00 | 6.13 | 13.87 |
| 165 | 10.31 | 6.70 | 13.93 |
| 170 | 10.63 | 7.31 | 13.94 |
| 175 | 10.94 | 7.99 | 13.89 |
| 180 | 11.25 | 8.73 | 13.77 |
| 185 | 11.56 | 9.61 | 13.51 |

### For 6 workers

| Total | X Bar (ave) | LCL | UCL |
|---|---|---|---|
| 70 | 2.92 | 0.00 | 7.37 |
| 75 | 3.13 | 0.00 | 7.69 |
| 80 | 3.33 | 0.00 | 7.99 |
| 85 | 3.54 | 0.00 | 8.28 |
| 90 | 3.75 | 0.00 | 8.57 |
| 95 | 3.96 | 0.00 | 8.84 |
| 100 | 4.17 | 0.00 | 9.11 |
| 105 | 4.38 | 0.00 | 9.38 |
| 110 | 4.58 | 0.00 | 9.63 |
| 115 | 4.79 | 0.00 | 9.88 |
| 120 | 5.00 | 0.00 | 10.12 |
| 125 | 5.21 | 0.06 | 10.36 |
| 130 | 5.42 | 0.25 | 10.59 |
| 135 | 5.63 | 0.44 | 10.81 |
| 140 | 5.83 | 0.64 | 11.03 |
| 145 | 6.04 | 0.85 | 11.24 |
| 150 | 6.25 | 1.06 | 11.44 |
| 155 | 6.46 | 1.28 | 11.64 |
| 160 | 6.67 | 1.50 | 11.83 |
| 165 | 6.88 | 1.73 | 12.02 |
| 170 | 7.08 | 1.97 | 12.19 |
| 175 | 7.29 | 2.22 | 12.37 |
| 180 | 7.50 | 2.47 | 12.53 |
| 185 | 7.71 | 2.73 | 12.69 |
| 190 | 7.92 | 2.99 | 12.84 |
| 195 | 8.13 | 3.27 | 12.98 |
| 200 | 8.33 | 3.55 | 13.12 |
| 205 | 8.54 | 3.83 | 13.25 |
| 210 | 8.75 | 4.13 | 13.37 |
| 215 | 8.96 | 4.44 | 13.48 |
| 220 | 9.17 | 4.75 | 13.58 |
| 225 | 9.38 | 5.08 | 13.67 |
| 230 | 9.58 | 5.42 | 13.75 |
| 235 | 9.79 | 5.76 | 13.82 |
| 240 | 10.00 | 6.13 | 13.87 |

The Lean Games and Simulations Book: Second Edition. Copyright © John Bicheno, 2014

# Motivation and Measures Game:
## Results Chart: Sample

| Name | Round 1 | Round 2 | Average so far | Round 3 | Average So far | Round 4 | Player Total | Player Average |
|---|---|---|---|---|---|---|---|---|
| Simon | 9 | 9 | 9 | 9 | 9 | 6 | 33 | 8.25 |
| John | 11 | 5 | 8 | 6 | 7.33 | 11 | 33 | 8.25 |
| Kim | 4 | 6 | 5 | 2 | 4 | 7 | 19 | 4.75 |
| Kate | 6 | 12 | 9 | 7 | 8.33 | 5 | 30 | 7.5 |
| Jane | 10 | 7 | 8.5 | 9 | 8.67 | 5 | 31 | 7.75 |
| Debbie | 8 | 4 | 6 | 7 | 6.33 | 10 | 29 | 7.25 |
| Round Total | 48 | 43 | | 40 | | 44 | Grand Total 175 | |
| Round Average | 8.0 | 7.17 | | 6.67 | | 7.33 | Overall Average 7.29 | |

# Targets and Measures Game: Results Chart

| Name | Round 1 | Round 2 | Average so far | Round 3 | Average So far | Round 4 | Player Total | Player Average |
|---|---|---|---|---|---|---|---|---|
|  |  |  |  |  |  |  |  |  |
|  |  |  |  |  |  |  |  |  |
|  |  |  |  |  |  |  |  |  |
|  |  |  |  |  |  |  |  |  |
|  |  |  |  |  |  |  |  |  |
|  |  |  |  |  |  |  |  |  |
| Round Total |  |  |  |  |  |  |  | Grand Total |
| Round Average |  |  |  |  |  |  |  | Overall Average |

The Lean Games and Simulations Book: Second Edition. Copyright © John Bicheno, 2014

A range of books and games is available from PICSE Books.

Please see www.picsie.co.uk and www.amazon.com

## Books

### The Lean Toolbox 4<sup>th</sup> Edition:

The Lean Toolbox 4$^{th}$ Edition:

The Essential Guide to Lean Transformation
John Bicheno and Matthias Holweg
Picsie Books 2009
A standard reference guide to Lean tools and thinking

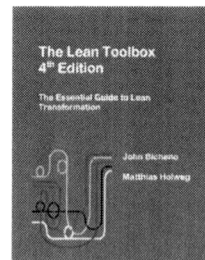

### The Service Systems Toolbox

John Bicheno
Picsie 2012

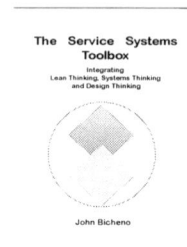

### Six Sigma and the Quality Toolbox

For service and manufacturing
John Bicheno and Philip Catherwood
Picsie 2005

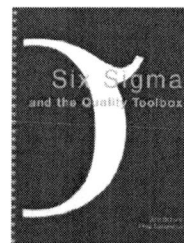

### Fishbone flow:

Integrating Lean, Six Sigma, TPM and TRIZ using fishbone diagrams
John Bicheno
Picsie 2006

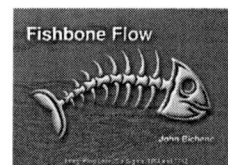

## Classroom games

### The Buckingham Lean Game

An ideal game to learn Lean manufacturing. This is a better game than the 'sticklebrick game' for scheduling. Unlike other games this simulation includes several products, changeover, quality and right size machines. It can be adapted to multiple environments. Over 600 copies of the game have been sold in Sweden alone.

### The Buckingham Service Operations Game

Built around an actual situation, this simulation is based on several Lean service concepts including service mapping, failure demand, 5S in service, and customer interactions.

Lightning Source UK Ltd.
Milton Keynes UK
UKOW01f0006080814

236539UK00001B/1/P